Contents

Contents

Construction Arbitrations

A Practical Guide

Second Edition

Vincent Powell-Smith
LLM, DLitt, FCIArb

John Sims
FRICS, FCIArb

and

Christopher Dancaster
DipICArb, FRICS, FCIArb

b

**Blackwell
Science**

© 1989 Ingramlight Properties Ltd and J H M Sims,
1998 The estate of Vincent Powell-Smith, J H M Sims
and C S Dancaster

Blackwell Science Ltd
Editorial Offices:
Osney Mead, Oxford OX2 0EL
25 John Street, London WC1N 2BL
23 Ainslie Place, Edinburgh EH3 6AJ
350 Main Street, Malden
  MA 02148 5018, USA
54 University Street, Carlton
  Victoria 3053, Australia
10, rue Casimir Delavigne
  75006 Paris, France

Other Editorial Offices:

Blackwell Wissenschafts-Verlag GmbH
Kurfürstendamm 57
10707 Berlin, Germany

Blackwell Science KK
MG Kodenmacho Building
7-10 Kodenmacho Nihombashi
Chuo-ku, Tokyo 104, Japan

The right of the Author to be identified as the Author
of this Work has been asserted in accordance with the
Copyright, Designs and Patents Act 1988.

First published 1989
by Legal Studies and Services Ltd
Second edition published 1998 by
Blackwell Science Ltd

Set in 10.5/12.5 pt Palatino
by DP Photosetting, Aylesbury, Bucks
Printed and bound in Great Britain by
MPG Books Ltd, Bodmin, Cornwall

The Blackwell Science logo is a trade mark of
Blackwell Science Ltd, registered at the United
Kingdom Trade Marks Registry

DISTRIBUTORS

Marston Book Services Ltd
PO Box 269
Abingdon
Oxon OX14 4YN
(*Orders:* Tel: 01235 465500
       Fax: 01235 465555)

USA
  Blackwell Science, Inc.
  Commerce Place
  350 Main Street
  Malden, MA 02148 5018
  (*Orders:* Tel: 800 759 6102
         781 388 8250
       Fax: 781 388 8255)

Canada
  Login Brothers Book Company
  324 Saulteaux Crescent
  Winnipeg, Manitoba R3J 3T2
  (*Orders:* Tel: 204 224-4068)

Australia
  Blackwell Science Pty Ltd
  54 University Street
  Carlton, Victoria 3053
  (*Orders:* Tel: 03 9347 0300
       Fax: 03 9347 5001)

A catalogue record for this title is available from the
British Library

ISBN 0-632-03992-2

Library of Congress
Cataloging-in-Publication Data
Powell-Smith, Vincent.
    Construction arbitrations: a practical guide/
  Vincent Powell-Smith, John Sims, and Christopher
  Dancaster. – 2nd ed.
      p. cm.
    Includes index.
    ISBN 0-632-03992-2 (pbk.)
    1. Construction contracts–Great Britain.
  2. Arbitration and award–Great Britain.
  I. Sims, John, 1929– . II. Dancaster,
  Christopher. III. Title.
  KD1641.P677 1998
  343.41′078624–dc21   98-12733
                                            CIP

For further information on Blackwell Science, visit
our website: www.blackwell-science.com

Preface to Second Edition

In the Preface to the first edition of this book the then authors said:

> 'This short book is intended to be a *vade mecum* for those involved or likely to be involved in a construction arbitration, and its emphasis is practical.'

The same can be said of this edition, although its scope has been extended to cover a wider field of construction arbitrations than before. Its basic intention is to provide practical advice, and to that end, following the weighty example of the Arbitration Act 1996, we have sought to avoid too much emphasis on legal terms and the over-use of reference to legal cases save where it is unavoidable.

A great deal has happened in the field of construction disputes in the nine years since the first edition was published. Not least, the coming into force on 31 January 1997 of the Arbitration Act 1996 bids fair to transform the whole practice of arbitration in construction disputes as well as in the many other fields of dispute to which it applies. Construction disputes have also been the subject of special legislative attention in the adjudication provisions in section 108 of the Housing Grants, Construction and Regeneration Act 1996. This book is about arbitration, and we have therefore not sought to comment on adjudication save where it is inevitable in examining the revised dispute resolution provisions in standard form contracts.

On the negative side there can be little doubt that arbitration as a means of resolving construction disputes has been increasingly the subject of adverse comment. Complaints of long delays, costs and occasional maverick decisions have led many advisers to the parties to construction contracts to recommend the exclusion of arbitration clauses in *ad hoc* contracts and by amendment to standard forms, often without thorough examination of the consequences. The removal of the court's discretion to refuse a stay of litigation where there is a valid arbitration clause in a contract by section 9 of the Arbitration Act 1996 has been seen by many as raising almost insuperable difficulties where disputes, as they so often do in the construction context, involve more than the two parties to a single contract. As will be apparent from the text of this book, we believe that these problems should be solved by sensible use by competent arbitrators – and party advisers – of the greatly increased powers now given to arbitrators by the Arbitration Act 1996, particularly the powers to control recalcitrant parties intent on delaying matters, and the use of sensible provisions in related main contracts, sub-contracts and consultants' contracts for dealing with disputes involving more than two parties. The production of the Construction Industry Model Arbitration Rules and their broad adoption by the Joint Contracts Tribunal in all their standard forms of contract and related sub-contracts and service contracts should serve almost completely to solve the problem of multi-party disputes in arbitration to the great gain of the industry, the associated professions and their clients.

However, the continued disillusionment with arbitration as a means of settling construction disputes is shown with brutal clarity by the decision of the Joint Contracts Tribunal to provide the option of litigation in many, if not all, of its standard form contracts. There can be little doubt that many building employers and their advisers will be tempted, at least in the short-term future, to choose the litigation option in the belief that it provides more certainty at less cost. With the greatest of respect to the courts, particularly the Official Referees' courts which have built up over the years enormous expertise in dealing with the technical as well as legal issues arising in construction disputes and which are in some cases undoubtedly the best forum for the determination of such disputes, we believe that those who now choose the litigation option will rapidly be brought to appreciate the advantages which arbitration provides, not least in the ultimate control of procedures, and above all timetabling, by the parties themselves where they can agree.

Both the present authors, like the late Vincent Powell-Smith, believe passionately that properly managed arbitration is, in most cases, by far the most effective, speedy and economical method of resolving disputes in the construction industry and its related professions. If we did not we would not be in the business as arbitrators. We have referred above to 'competent' arbitrators. The choice of a truly competent arbitrator properly able to run the arbitration in accordance with both the spirit and the letter of the Arbitration Act 1996 with its emphasis on fairness, speed and economy in the management of arbitrations, is crucial to the success of any arbitration. We hope that this new edition of this book will serve, if only in a very small way, to increase the cadre of competent arbitrators available to the industry, professionals and clients, to the ultimate benefit of them all.

Stop press

While this book was in the final stages of printing the House of Lords handed down its decision in the case of *Beaufort Developments (NI) Limited* v. *Gilbert Ash NI Limited and Others* (20 May 1998 – officially unreported at the time of writing but available on the House of Lords Internet site). This decision reverses that of the Court of Appeal in *Northern Regional Health Authority* v. *Derek Crouch Construction Co Ltd* (1984) 26 BLR 1.

Following this new decision there is no bar to the courts opening up, reviewing and revising any certificate, opinion, decision or notice given by an architect or engineer under a contract where such a certificate, opinion, decision or notice is arguably not in accordance with the requirements of the contract. References in the following text to the effect of the *Crouch* decision (see, for instance, page 5) should therefore be ignored. It is probably also now unnecessary for the JCT to include in its new clause 41C, providing for the resolution of disputes by litigation, a special power for the court to do what the House of Lords has now said it can do anyway (see page 295). Section 100 of the Courts and Legal Services Act which allows the High Court to exercise such powers if all parties to an arbitration agreement agree, is also now probably rendered unnecessary.

It may still be argued that the powers of an arbitrator to re-examine certificates etc are wider than those of a court because his power does not depend upon any possible breach of contract. A certificate may arguably be compliance with the requirements of the contract, but there may still be a dispute as to its 'rightness' simply as a matter of professional opinion. No two quantity surveyors, for instance, will agree absolutely on a valuation of work for interim payment. Both opinions may be contractually right so that the implementation of either may not entail any breach of the contract conditions, and there would then be nothing for the court to open up, review or revise. A technical arbitrator, however, may prefer one view over the other and substitute his view for that of the contract quantity surveyor. However, it is equally arguable that, since a certificate is a contractual document establishing the parties' rights and obligations under the contract, any dispute as to its correctness must, by definition, be a matter of contract which a court may now examine. There is still much scope for argument on the point both inside and outside the courts, and the authors await further developments with interest.

John Sims
Christopher Dancaster
Chelsea/Milton Keynes, June 1998

Vincent Powell-Smith

I cannot let the publication of this second edition of a book which I originally wrote with the late Vincent Powell-Smith pass without paying my own tribute to that extraordinary man with whom I had the privilege and pleasure of collaborating so many times.

Vincent was a man of great force of character, and like most such men had both great faults and great virtues. His faults, such as they were, were a part of his personality, and this is not the place to dwell on them; they have died with him. His personal virtues were many and included great personal kindness and generosity, certainly to me.

His virtues as a lawyer and as a writer were apparent from his many books and other publications, some with me but many more with other collaborators and on his own, covering a staggeringly wide field, not only of the law but also of other unexpected things such as horse management and genealogy. His knowledge of the law was encyclopaedic. In our collaborations he was able, without hesitation, to go to the many volumes of law reports weighing down his shelves and go unerringly to the most obscure of cases to support the point under discussion.

He was also the easiest and most co-operative of collaborators. While we had furious rows on many matters, usually political where our views widely diverged, when writing together we almost invariably agreed, and even where we initially disagreed we were almost always able to reach a consensus view, sometimes leaning towards his initial opinion and sometimes towards mine. On the very few occasions when we were unable to reach final agreement the disagreement was always amicable and we were able to state our differing views openly and frankly in the text.

His death has left a Vincent-shaped hole in the world which no-one else will quite be able to fill. I miss him, and wish he were still alive to have another of our political rows which sometimes ended in one of us – usually me – storming out of the room and slamming the door but which had invariably disappeared within the next five minutes when we would resume drinking our Scotches and amiably discussing something else. Wherever he is now, I wish him well.

John Sims
Chelsea
June 1998

Chapter One
What is Arbitration?

1.1 General introduction

1.1.1 Definition

Arbitration is a process whereby the parties to a dispute agree to have it settled by an independent third party and to be bound by the decision he makes. This agreement may be entered into after the dispute has arisen or it may be included within a contract by way of a clause which refers any future dispute which might arise out of that contract to arbitration. The third party may be chosen by agreement between the parties themselves or he may be appointed by a person named in the contract to carry out that function. Commonly in the construction industries such appointments are made by the Presidents of the Royal Institute of British Architects, the Royal Institution of Chartered Surveyors, the Institution of Civil Engineers or the Chartered Institute of Arbitrators.

There are three essential elements of arbitration, namely:

- the existence of a dispute between the parties,
- an agreement between them to refer it to arbitration, and
- both parties agreeing to be bound by the decision of the arbitrator.

Arbitration will be selected by parties as the agreed contractual method of resolving disputes for various reasons, among which will be:

- they consider that it is appropriate for a dispute between them to be resolved by someone who is familiar with the technicalities of their field,
- the familiarity of the arbitrator with the technical aspects of their dispute will lead to economies of time and cost in resolving the dispute,
- they require the privacy which arbitration provides,

- arbitration can be completed without the long delays inherent in the court process.

The arbitrator is not a judge sitting in a court provided and financed by the state, he is an independent professional person who, in addition to his technical knowledge of the matters in dispute should generally have some considerable experience of the arbitral process.

1.1.2 Comparison between arbitration and litigation

Where litigation is used the process can be slow in itself due to the procedures that have to be followed and it can take considerable time for a dispute to get before the court. The limited capacity of the courts means that if this were to be the only method of formal dispute resolution available to the parties to a dispute the whole process would inexorably grind to a halt. It is bound by formal procedures and in particular by the Rules of the Supreme Court. It is always an adversarial process in common law jurisdictions. This can lead to a considerable outlay in costs of preparatory work and the detailed procedures that have to be followed. The great advantage is that the services of the judge and the accommodation for the trial are provided by the state. Judges are also the possessors of far greater powers than arbitrators. Legal aid is also available in certain circumstances in litigation.

Arbitration, on the other hand, where both parties to a dispute want it to be resolved and to get on with their respective businesses, provides an infinite variety of procedures which can be adopted. It is possible in smaller cases to dispense with formal procedures almost entirely and by means of documentary submissions alone reach a position that the arbitrator can make his award at minimal cost to the parties. Arbitration can, however, suffer all the same problems of delay and cost as do the courts especially where there has been a unilateral application to a person named in the contract and the other party is, for reasons best known to itself, seeking to delay matters. The Arbitration Act 1996 (referred to from now on as 'the 1996 Act') gives the arbitrator rather more powers than he has had under the previous regime and encourages him to use them robustly in instances of this nature.

1.1.3 Comparison between arbitration and mediation/conciliation

A majority of disputes have always been settled by negotiation. Where negotiation fails, conciliation and mediation can be used instead. Conciliation and mediation are procedures whereby an independent, impartial person assists the parties to reach an agreed settlement of their dispute. The difference between the two processes is that there is no obligation on the mediator other than to facilitate the possibility of settlement whereas the conciliator may, if necessary, offer a suggested solution to the dispute for the parties to consider. The process in either case is voluntary. If the parties reach a settlement as a result of the process, once recorded it becomes a contract between them which is enforceable.

The great disadvantage of mediation and conciliation is that they do not necessarily produce a resolution to the dispute. If they do not resolve the dispute the parties then have to resort to arbitration or litigation. Even if a settlement is reached the resultant agreement is only enforceable in contract and may be disputed if one party does not wish to honour it.

There is currently a trend towards the development of dispute resolution procedures within contracts which include a formalised conciliation stage as a precursor to arbitration. These include the Institution of Civil Engineers' Design and Construct Contract (1992) and the Chartered Institute of Arbitrators' Consumer Dispute Resolution Scheme. Other contracts published by the Institution of Civil Engineers, the ICE Sixth Edition 1991 and the ICE Minor Works Contract (1988) include a conciliation procedure but in these contracts it is optional, as is the conciliation procedure included in the Federation of Civil Engineering Contractors' Sub-contract (September 1991).

1.1.4 Comparison between arbitration and adjudication

Arbitration and adjudication are similar in one major respect. They both seek to establish the contractual rights, duties and obligations of the parties. Both the arbitrator and the adjudicator have in principle to act in a way that complies with the concept of fairness.

An adjudicator can be named in the contract if so desired by the parties or once a dispute has arisen he can be agreed by the parties or failing this appointed by a third party.

Adjudication, with the enactment of the Housing Grants, Construction and Regeneration Act 1996 (at the time of writing awaiting the Scheme for Construction Contracts before it comes into force) becomes a statutory right to any party to a construction contract. It is not therefore a consensual process as is the case with arbitration.

Adjudication has been described as 'rough justice' and 'a quick and dirty fix' and there is no doubt that the aspirations of those championing adjudication are that it is better to obtain the decision of a respected professional man on the basis of limited information in a very short period of time than to spend months if not years reaching a not dissimilar result at vast expense. There are possible problems with the application of such a process to major disputes in that the obligation to pay large sums of money as a result of a possibly superficial examination of the facts may well not go down too well with the paying party. The hoped-for alternative is that this result will be accepted and the dispute thus resolved without the expenditure of vast sums of money in litigation or arbitration costs.

The ideal is, of course, to restrict the adjudication to discrete issues before there has been the opportunity for large amounts of money to become the subject of dispute. The opportunity to seek the decision of an adjudicator at any time may well mean that each aspect of what might finally turn out to be a complex dispute has been examined and resolved as the work proceeds.

A useful example is the typical construction dispute which revolves around a multitude of events which may or may not have caused delay or disruption. If each of these events can be examined by an adjudicator and a decision reached as to the effect of each of them the opportunity for the development of a complex claim is very much reduced.

The principal difference between arbitration and adjudication, other than the time scale for making the decision, is the aspect of finality. Whilst a decision reached by an adjudicator is as binding on the parties as an arbitrator's award, the dispute that has been the subject of adjudication proceedings can be the subject of completely fresh arbitration or litigation. The adjudicator's decision remains in place and must be honoured whilst the arbitration or litigation reconsiders the dispute from first principles.

It may be thought from the experience of the number of arbitrators' awards in construction matters that have been the subject of subsequent court proceedings that arbitration, at least in the construction industry, does not even have the benefit of finality. This however is not the norm. The awards that get into the courts are a minority, however well publicised; most construction arbitrators' awards are accepted by the parties and no further proceedings result.

An adjudicator's decision may, as a result of the existence of an arbitration clause in the contract, have to be enforced by an arbitrator. This is a matter of some significance and is considered in Chapter 3.

1.1.5   Comparison between arbitration and expert determination

An 'expert' can be appointed to determine a dispute between two parties in a similar way to an arbitrator, a mediator or an adjudicator. He can be agreed between the parties or he can be appointed by a third party. He can also, as with an adjudicator, be named in the contract.

Whilst adjudicators and arbitrators may use their own expertise if so agreed by the parties, this is not always the case. Expert determination is distinguished by the fact that it specifically envisages that the expert will use his own expertise.

The procedure is usually a simple one and based upon the wording of the specific contract. The main difference between expert determination and arbitration is that the former is subject to little or no control by the courts. Experts are appointed entirely on a contractual basis and may be liable for negligence. Unless an expert gives reasons, his determination is unlikely to be appealable, see *Conoco and Others* v. *Phillips Petroleum Company and Others* [1996] CILL 1204. The decision that the expert makes has a completely different status from that of a judge or arbitrator. It is only enforceable as a contractual provision.

1.1.6   Advantages of arbitration over other forms of dispute resolution

The principal advantages of arbitration are that it produces a legally binding decision by means of a process that is, if desired, totally within the control of the parties and that it can be as flexible as necessary to suit their requirements.

One of the perceived advantages of an arbitration clause in a construction contract has for many years been the power of the arbitrator to open up, review and revise any certificate, opinion, decision, requirement or notice. The Court of Appeal in *Northern Regional Health Authority* v. *Derek Crouch Construction Co Ltd* (1984) 26 BLR 1 found that the courts did not possess these powers in the absence of a specific agreement between the parties to that effect. The power vested in the arbitrator is therefore of considerable assistance to a contractor seeking to vary an architect's decision relating, for example, to an extension of time which he would not be able to get the court to do by virtue of *Crouch*.

The courts have nibbled away at the inability to review an architect's decision over the years since *Crouch* was decided. It has always been possible for the court to overturn an architect's decision if that decision was made in bad faith or in excess of his powers. In a recent decision, *John Barker Construction* v. *London Portman Hotel* (1996) 12 Const LJ 277 the court held that it can interfere with an architect's decision if the contractual machinery has broken down or if the architect has not acted lawfully or fairly. The court found that the architect's extension of time was fundamentally flawed and, in view of the lapse of time since the original decision was made, the contractual machinery had broken down to such an extent that it would not be practicable or just for the matter to be remitted to the architect for redetermination. As a result of this decision the court went on to determine the length of a fair and reasonable extension of time.

1.2   *The statutory framework*

English arbitrations are governed by the 1996 Act which came into force on 31st January 1997. Appendix 1 gives the text of this statute.

Also of relevance to English arbitration is the 'Report on the Arbitration Bill' produced by the Departmental Advisory Committee on Arbitration Law ('the DAC') in February 1996 and the Supplementary Report produced in January 1997. The DAC, under the chairmanship of the Rt Hon Lord Justice Saville (as he then was), was responsible for the drafting of the Arbitration Bill which ultimately became the 1996 Act. This Report sets out in detail the thinking behind the provisions included in the 1996 Act and for those who are interested is extremely useful reading. It is however beyond the scope of this practical guide.

The 1996 Act is divided into four parts:

Part I   'Arbitration pursuant to an arbitration agreement'
Part II   'Other provisions relating to arbitration'
Part III   'Recognition and enforcement of certain foreign awards'
Part IV   'General provisions'

The 1996 Act as a whole contains 110 sections and Part I with 84 sections is the most extensive and relevant to the conduct of a construction arbitration.

Part I

This is set out in an order reflecting that which an arbitration generally follows and, after a number of introductory clauses set out in sections 1–5, covers such matters as the arbitration agreement in sections 6–8, the commencement of the arbitral proceedings in sections 12–16, the arbitral proceedings in sections 33–41 and the award in sections 46–58.

Part II

As enacted this included, in sections 85–8, provisions covering domestic arbitration agreements. These provisions were contrary to European law and were not brought into force with the rest of the 1996 Act. This does affect the approach of the court to applications to stay court proceedings to arbitration, of which more later. The remainder of this Part covers consumer arbitration agreements and such matters as the appointment of judges as arbitrators and statutory arbitrations none of which are of any great relevance for the purposes of this book.

Part III

This deals with foreign awards and is also beyond the ambit of this book.

Part IV

This includes various general provisions.

There are three particular sections of the 1996 Act which are of such importance to the process that it is worth setting them out in full here. These are sections 1, 33 and 40.

Section 1 of the 1996 Act

This sets out the general principles upon which the 1996 Act is based. It is couched in terms that any ambiguity elsewhere in the 1996 Act will be overridden by the principles set out in section 1. This section is as follows:

'The provisions of this Part are founded on the following principles, and shall be construed accordingly —
(a) the object of arbitration is to obtain the fair resolution of disputes by an impartial tribunal without unnecessary delay or expense;

(b)   the parties should be free to agree how their disputes are resolved, subject only to such safeguards as are necessary in the public interest;

(c)   in matters governed by this Part the court should not intervene except as provided by this Part.'

This section makes the object of arbitration quite clear: 'to obtain the fair resolution of disputes by an impartial tribunal without unnecessary delay or expense'. It also clarifies the fact that it is the parties' arbitration and that they must have the opportunity to decide the procedure by which their dispute is to be resolved. In practice this is not always possible as a dispute on substantive issues very often extends into a disagreement on the proper procedural approach to the arbitration itself. Other sections of the 1996 Act deal with the powers of the arbitrator where the parties do not agree and these will be looked at in detail later.

Section 33 of the 1996 Act

This sets out the general duty of the tribunal.

'(1)   The tribunal shall —

(a)   act fairly and impartially as between the parties, giving each party a reasonable opportunity of putting his case and dealing with that of his opponent, and

(b)   adopt procedures suitable to the circumstances of the particular case, avoiding unnecessary delay or expense, so as to provide a fair means for the resolution of the matters falling to be determined.

(2)   The tribunal shall comply with that general duty in conducting the arbitral proceedings, in its decisions on matters of procedure and evidence and in the exercise of all other powers conferred on it.'

The DAC Report states that it is the intention of the 1996 Act to give maximum flexibility to arbitrators in their handling of the dispute. This section sets out that intention quite clearly and offers the opportunity for the avoidance of arbitration procedures which resemble those of the court.

Section 40 of the 1996 Act

This sets out the general duty of the parties and seeks to avoid the possibility of interference with the process by a reluctant party.

'(1) The parties shall do all things necessary for the proper and expeditious conduct of the arbitral proceedings.

(2) This includes —

    (a) complying without delay with any determination of the tribunal as to procedural or evidential matters, or with any order or directions of the tribunal, and

    (b) where appropriate, taking without delay any necessary steps to obtain a decision of the court on a preliminary question of jurisdiction or law.'

1.3 Arbitration agreements

Although it is quite sufficient for the disputing parties to state orally: 'We agree that the disputes between us be referred to arbitration', this, without more, could lead to difficulties. Although an oral arbitration agreement is quite valid, the resulting arbitration will not be governed by the 1996 Act. This could lead to difficulties in enforcing the award. In practice, therefore, it is essential that the arbitration agreement be in writing. Section 5 of the 1996 Act sets out the requirements relating to agreements in writing. This section is drawn very widely and, in addition to those agreements that might be in normal written form, it includes agreements that are evidenced in writing. This means, for example, that a minute of a meeting recording that the parties to a contract agreed in discussion that disputes would be resolved by arbitration fulfils the requirements of a written arbitration agreement.

If the arbitration agreement is in writing, it will be subject to the 1996 Act, even if no express reference is made to it, and as a result of section 15(2), the reference will be deemed to be to a single arbitrator unless the agreement states otherwise.

In those unusual cases where a dispute arises from a contract which does not contain an arbitration clause, the parties are free to enter into an *ad hoc* arbitration agreement in respect of the disputes between them. This is of relevance in the construction industry where, for example, there is a contract for alteration works between a builder and an employer, and the contract is not in any of the standard forms. This situation is far more common than is generally supposed. In such a case, arbitration may provide a sensible method of settling disputes between the employer and the contractor, and if those parties decide that arbitration is appropriate they will have to enter into a suitable arbitration agreement.

Section 3 of the 1996 Act allows the parties to designate the 'Seat of the Arbitration'. Strictly speaking this is only really important to

Figure 1.1 is an example of an ad hoc arbitration agreement.

FIGURE 1.1 – AD HOC ARBITRATION AGREEMENT

In the Matter of the Arbitration Act 1996

and

In the Matter of an Arbitration

between

[*Name of Claimant*]   CLAIMANT

and

[Name of Respondent]   RESPONDENT

AGREEMENT TO REFER

A dispute has arisen between the parties named above under an Agreement between them dated [*date*] relating to certain works executed by the Claimant at [*address*]. The parties hereby agree to refer all matters in dispute or difference between them arising out of or in connection with the said Agreement to the arbitration and final decision of [*name and qualifications of arbitrator*] of [*arbitrator's address*] ALTERNATIVELY [*a person to be appointed by the President or a Vice-President of the Chartered Institute of Arbitrators upon the joint application of the parties*]

The seat of the arbitration shall be England.

Dated. . . . .

Signed . . . .   for and on behalf of the Claimant

Signed . . . .   for and on behalf of the Respondent

---

help define the applicable law in the case of international arbitrations. There is however in section 52 of the 1996 Act a requirement that any award shall state the seat of the arbitration. In the interests of certainty an *ad hoc* agreement will usefully include this information.

All the widely-used standard form building and civil engineering contracts contain an arbitration clause by which the parties to the contract agree to refer disputes arising between them to arbitration. Agreements of this kind are called 'agreements to refer'. An 'agreement to refer' incorporated into a contract is enforceable as a

10

term of that contract and, indeed, if for some reason the contract is brought to an end, an arbitration agreement would remain separately enforceable as an arbitration agreement by section 7 of the 1996 Act.

1.3.1 Staying legal proceedings where there is an arbitration agreement

Where the contract contains an 'agreement to refer' the courts will in principle uphold arbitration as the method of dealing with disputes. The mere existence of an arbitration clause does not generally prevent one or other party commencing proceedings in the courts, but section 9 of the 1996 Act empowers the court to stay the litigation proceedings, provided that the party wishing to have the dispute referred to arbitration applies to the court but not 'before taking the appropriate procedural step (if any) to acknowledge the legal proceedings against him or after he has taken any step in those proceedings to answer the substantive claim' (section 9(3)).

The decision not to bring sections 85–8 of the 1996 Act into force, as mentioned above, does create one particular problem and that relates to the dispute that may not be appropriate for resolution by arbitration. In the past the court has decided that in cases where there are disputes on related contracts and there is the possibility of two different arbitrators making differing decisions on effectively the same facts it is most appropriate that the two disputes be dealt with as one by the court and the agreement to arbitrate is ignored.

The flexibility afforded to the court by section 86 is not there as a result of these clauses not being brought into force. Parties entering into contracts which are related to one another such as main and sub-contracts will need to make provision for the joinder of these disputes and include appropriate machinery within their contracts to ensure that this is made possible. Provisions which are intended to fulfill these requirements are included in many of the standard form contracts. The Act itself in section 9 empowers the parties to an arbitration to agree the consolidation of two (or more) arbitral proceedings but this is almost certainly only likely to occur after the dispute has arisen. There may at this time be a certain reluctance on the part of one party to reach such an agreement in a situation where there are no joinder provisions in the contract. The Construction Industry Model Arbitration Rules ('CIMAR') (considered in more detail later) seek to avoid this problem by removing the decision from the parties. These rules include a provision whereby the parties agree at the time of entering into the contract to empower an arbitrator to consolidate arbitral proceedings where he considers it to be appropriate. This will of course only apply where rules containing this provision apply to both contracts and is a strong argument for the adoption of a single standard set of Rules on all construction contracts.

It is, in fact, possible to make arbitration a precondition to litigation by means of an appropriately worded clause. Such a provision is known as a *Scott* v. *Avery* clause after the case of that name decided by the House of Lords in the last century: see (1856) 5 HL Cas 811. In that case, which involved an insurance contract, it was provided that the insurer was not to maintain an action at law in respect of the policy until arbitrators had given a decision, and the effectiveness of the provision was upheld by the House of Lords.

Equally, it is possible to provide in a contract that unless arbitration is commenced within a specified period, all claims 'shall be deemed to be waived and absolutely barred'. The validity of a provision worded in this way was upheld by the House of Lords in *Atlantic Shipping and Trading Co* v. *Louis Dreyfus & Co* [1922] AC 250, though section 12 of the 1996 Act gives the courts the power to extend the period in cases of hardship.

Clauses of these types are seldom found in construction contracts and none of the standard form arbitration clauses fall into either category. Such clauses are, however, common in the maritime and insurance fields.

1.4 Types of arbitration

One of the great advantages of arbitration is its flexibility. How the arbitration is to be conducted is very much a matter for the parties to decide, although the aim should always be to allow the disputants to obtain a just decision.

The simplest type of arbitration is what a former Master of the Rolls (Lord Donaldson) has referred to as the 'look-sniff' arbitration. This is common in the grain trade and simply consists of an expert arbitrator examining samples of grain to determine whether they are up to contract standard. Some would argue that strictly speaking this is not arbitration at all but is rather analogous to valuation. Nevertheless, it seems to be generally accepted as a species of arbitration.

There is no reason why a similar simple procedure could not be used in the building industry where the dispute relates to quality of work, for example, or whether works have, in fact, been 'practically completed'. This method could certainly be used in a 'small works' type arbitration where – from a monetary point of view – comparatively small sums are at stake.

Normally, in the construction industry the simplest type of arbitration is what is known as 'documents only' arbitration, in which there is no hearing before the arbitrator who reaches his decision on the basis of the parties' written submissions supported by documentary evidence. In some cases there may be an inspection of the work by the arbitrator in the presence of the parties or their representatives. This is a method often used, for instance, in arbitrations under the National House-Building Council Buildmark Scheme. The Construction Industry Model Arbitration Rules and the ICE Arbitration Procedure both include such a procedure.

Many common contractual disputes can easily be settled on a documents only basis, and in an appropriate case the arbitrator may suggest this method of approach. Parties, particularly when they are private individuals, do however like to feel that they have had their contentions heard and properly considered. Some private individuals also have difficulty in presenting their case adequately in written form. In these instances, and they are usually disputes of a limited financial extent, it is not unusual for a primarily documents only arbitration to include a short hearing where each party can 'have his say'. It is an accepted fact that unrepresented parties can be frightened by too much formality and this hearing will often be very informal and be described as a meeting.

In construction arbitrations of any complexity it is usual for there to be a hearing for the purpose of oral evidence being given on a relatively formal basis.

1.5 Standard arbitration clauses

Although most standard form building contracts contain an arbitration provision, the clauses differ in their wording and it is always necessary to examine each clause to ascertain its scope and the powers given to the arbitrator.

The principal points to ascertain from an examination of an arbitration clause are as follows:

- What effect does the clause have on the arbitrator's powers? Does it limit them or give him more than the 1996 Act does?
- Are there any rules incorporated?
- What restrictions, if any, are placed upon his jurisdiction? (This

may be further restricted by the notice of dispute which may only refer certain specific matters to the arbitrator for his decision.)

- Are any matters excluded from arbitration in any event because they are decisions that are stated by the contract to be final and conclusive?
- Is the wording of the clause sufficiently wide to give the arbitrator the power to deal with torts such as misrepresentation or to rectify the contract? *Ashville Investments Ltd* v. *Elmer Contractors Ltd* [1989] QB 488; [1988] 2 All ER 577, 37 BLR 55, CA construed the words 'in connection with' in JCT 80 Article 5 to extend the arbitrator's jurisdiction to cover misrepresentation.
- Is there any specific time requirement for the commencement of arbitration? Is it, for example, precluded until after practical completion?
- Is there any requirement that the architect, for example, should give a written decision on a matter before it can be submitted to an arbitrator?
- What are the provisions, if any, for the joining of disputes or the consolidation of proceedings?
- Is there any precondition to arbitration? For example, does the contract require that the dispute has been placed before an adjudicator or have been subject to conciliation proceedings?
- Are there any time limits on the proceedings? For example, must the statements of case and defence be prepared and does the arbitrator have to make his award within a specified period?

Some arbitration clauses include a provision that overcomes an obstacle that statute places in the way of any party seeking to appeal an arbitrator's award. Section 69(2) of the 1996 Act requires that the parties agree that an appeal on a point of law may be made before it can be brought before the court or that the court itself gives leave. In the absence of agreement between the parties the court will apply stringent criteria before allowing leave to appeal. A dissatisfied party is unlikely to get the other party to agree to an appeal after that other party has won on a point of law in an award. The inclusion of an agreement in the arbitration clause to allow an appeal on a point of law pre-empts the need for this agreement after the award has been made. This provision, which exists in a number of standard construction contracts, has resulted in both benefit and detriment to the process of dispute resolution in the construction industry. The principal benefit has been that a body of law has developed on construction cases that might otherwise not have been the case. The disadvantage is a reduction in the finality of the arbitral process and an increase in the cost of dispute resolution in the construction industry.

1.6 Arbitration rules

Arbitration rules are prepared to accompany various contracts and extracts from the Construction Industry Model Arbitration Rules (CIMAR) and the Institution of Civil Engineers' Arbitration Procedure (the ICE Procedure) are set out in the Appendices. These rules codify the powers and duties of the parties and the arbitrator setting out the agreement of the parties in that respect. The adoption of a set of arbitration rules in a contract avoids the necessity for the parties to consider the powers and duties that the arbitrator should have once the dispute has arisen and leaves the way clear for the arbitration to proceed at an early date.

CIMAR and the ICE Procedure illustrate two fundamentally different ways of approaching arbitration rules.

CIMAR rely heavily on the use of extracts from the 1996 Act adopting a process of amplification and essentially filling in certain gaps. They reproduce the relevant sections of the 1996 Act throughout and are very clearly designed for use in arbitrations where the 1996 Act applies, that is in England and Wales and Northern Ireland. The ICE Procedure on the other hand has the express intention of producing a set of rules which complies with the 1996 Act but does not rely on it to operate. It is intended to stand alone. It allows for the conduct of the arbitration by reference to the ICE Procedure alone without the need to refer back to the 1996 Act.

CIMAR have been produced for situations where the 1996 Act applies and are not intended for use wider afield. During their development they have been the subject of an extensive consultation with a wide spectrum of the construction industry. The ICE Procedure on the other hand envisages use in both the domestic and the international arenas. The authors of the ICE Procedure also consider that smaller arbitrations will benefit from the procedure being all inclusive without the need to refer back to the Act.

As noted previously, the Act allows the parties virtually total freedom to decide the way that their dispute is to be resolved. It is of course a fact that the majority of people who enter into construction contracts never allow their disputes to reach a situation where a third party dispute resolution procedure is necessary. As a result they have no experience to guide them as to the most appropriate way forward when faced with a myriad of decisions that have to be made. The aim of arbitration rules is to provide a standard means whereby these decisions with regard to the conduct of the arbitration are made at the outset of the contract.

CIMAR are reproduced in Appendix 2 and the ICE Procedure in Appendix 3.

1.7   The powers and duties of the parties

1.7.1   Powers

The principal power that the parties to an arbitration have is that given by section 1(b) of the 1996 Act to agree any procedure that they like for their arbitration apart from something which is contrary to the public interest.

There are a number of ways in which the parties to an arbitration can decide the procedure that they are going to adopt. The arbitration clause in the contract, be it in a standard form or bespoke, may itself address that procedure. As noted above this arbitration clause may refer to a set of arbitration rules which usually include a number of different procedures covering the various circumstances that may apply to the arbitration. The parties could sit down once the dispute has arisen and agree the procedure, although this is somewhat unlikely unless each party understands, or is represented by someone who understands, the benefits of so doing. The arbitrator himself may suggest the detail of the procedure and the parties may accept his suggestions. The parties could even agree a procedure that is contrary to the public interest but were they to do this, the resolution of their dispute would not be regulated by the 1996 Act or, if necessary, supported by the courts.

If no agreement is reached the procedure for the arbitration will be governed by the various provisions in the 1996 Act that apply in the absence of agreement to the contrary.

There are also a number of other powers vested in the parties by the 1996 Act. See Appendix 1 Table 1(b)

1.7.2   Duties

The 1996 Act places relatively few duties on the parties. Those duties that the parties do have are mainly the basic requirements for the success of the process. Not the least of these is section 40 which,

16

although quoted in 1.2 above, is of sufficient importance to repeat here and which requires that

'(1)   The parties shall do all things necessary for the proper and expeditious conduct of the arbitral proceedings.
(2)   This includes
    (a)   complying without delay with any determination of the tribunal as to procedural or evidential matters, or with any order or directions of the tribunal, and
    (b)   where appropriate, taking without delay any necessary steps to obtain a decision of the court on a preliminary question of jurisdiction or law.'

See also Appendix 1 Table 1(c).

1.7.3   Powers and duties of the parties under arbitration rules

As stated in 1.6, at the time of writing there are two sets of arbitration rules which are available for use in connection with the 1996 Act. These are CIMAR and the ICE Procedure.

By using CIMAR the parties agree that once an arbitrator has been appointed they will not amend the rules or impose procedures that conflict with them without the agreement of the arbitrator. One of the most important aspects of CIMAR is that they empower a party unilaterally to refer a further dispute under the same contract to be dealt with under the same proceedings. The two disputes *must* be consolidated if this is done before an arbitrator is appointed. If the arbitrator is already appointed he has the option to consolidate or not. This must be right as it could be totally inappropriate to delay the resolution of the first dispute if the further application is made late on in the proceedings. CIMAR also deal with the situation where there are two or more arbitral proceedings involving some common issue, providing a procedure whereby the parties to those arbitrations, be they two or more, can agree to consolidate the proceedings.

CIMAR also require the arbitrator to decide whether the procedure for the arbitration is to be a 'Full Procedure' culminating in a hearing, a 'Documents Only' procedure or a 'Short Hearing' procedure. This decision is made after the parties have made their own submissions and proposals. None of these procedures contain specific time requirements, except for a requirement for the arbitrator to make his award within one month in the documents only and short hearing procedures. Provisions similar to those in the 1988 JCT Arbitration Rules for the dismissal of a claim for failure without good reason to comply with time limits are not included. Section 41(5), (6) and (7) of the 1996 Act provides the arbitrator with powers to deal with dilatory parties. There is no doubt that it is now incumbent on arbitrators under section 33 of the 1996 Act to set sensible time scales and to require that these be complied with.

The principal modifications that CIMAR make to the parties' powers and duties are set out in Appendix 2 Table 2(a) and Appendix 2 Table 2(b).

The effect of the ICE Procedure on the parties' powers and duties is very similar to CIMAR. The parties have the option to choose a 'Short Procedure' which is essentially a documents only process with the possibility of limited oral submissions by the parties or questions by the arbitrator. They also have the option to choose a 'Special Procedure for Experts' which is similar to the short procedure but the hearing is limited to the arbitrator meeting the parties' respective experts. The authors of the ICE Procedure consider that it is appropriate for the short procedure to include time limits allowing 30 days for the preparation of the claimant's case and the respondent 30 days to reply. There are no specific sanctions for failure to comply with these time limits and section 41 of the 1996 Act will apply in this instance.

The principal modifications that the ICE Procedure makes to the parties' powers and duties are set out in Appendix 3 Table 3(a) and Appendix 3 Table 3(b).

1.8   The powers and duties of the arbitrator

1.8.1   Powers

The main thing to remember when considering the arbitrator's powers is the effect of section 1(b) of the 1996 Act which gives the parties complete freedom to agree how their dispute is to be resolved. Taken to the extreme this means that the arbitrator could enter into an arbitration with no power whatsoever to influence the conduct of the proceedings. The analogy of the referee without the right to use his whistle is irresistible in these circumstances and it is inconceivable that any party to an arbitration or any arbitrator would accept such a situation.

There are many powers set out in the 1996 Act for the arbitrator to use. These are in most cases only available to the arbitrator as a fall-back. If the parties agree something different the arbitrator will be constrained by that agreement and will have to conduct the arbitration by the parties' rules. The powers set out in the 1996 Act are there for good reason and it is suggested that the majority of arbitrations will be conducted with the arbitrator enjoying most if not all of the powers which it includes.

The arbitrator's powers come in three categories in the 1996 Act:

- those which the parties cannot take away;
- those that the arbitrator has unless the parties agree otherwise;
- those that the arbitrator may only be given by the agreement of the parties.

These are set out in Appendix 1, Tables 1(d)1, 1(d)2 and 1(d)3.

1.8.2 Duties

In addition to the powers given to him by the 1996 Act the arbitrator also has certain duties. These duties are encapsulated in section 33. Again this is set out in 1.2 above but must be considered to be of sufficient importance to repeat here. By this section the arbitrator must

'(a) act fairly and impartially as between the parties, giving each party a reasonable opportunity of putting his case and dealing with that of his opponent, and

(b) adopt procedures suitable to the circumstances of the particular case, avoiding unnecessary delay or expense, so as to provide a fair means for the resolution of the matters falling to be determined.'

The 1996 Act imposes a great number of duties on the arbitrator. These are set out in Appendix 1 Table 1(e). Some of these are mandatory, others are subject to the possibility that the parties may agree that they do not apply.

1.8.3 Powers and duties of the arbitrator under arbitration rules

As with the parties the application of a set of arbitration rules to the reference will affect the powers and duties of the arbitrator.

Neither CIMAR nor the ICE Procedure make any agreements

between the parties that remove any of the powers that the arbitrator has under the 1996 Act in the absence of such agreements. They both extend the powers of the arbitrator by allowing him to order consolidation (section 35) and provisional relief (section 39), powers that he otherwise only has if specifically agreed by the parties. There are also a number of provisions which amplify and clarify the arbitrator's powers and duties, principally procedural, and these are set out in Appendix 2 Tables 2(c) and 2(d) and Appendix 3 Tables 3(c) and 3(d).

One difference in approach is that CIMAR make the assumption that there will be a preliminary meeting unless the arbitrator considers it unnecessary. The ICE Procedure on the other hand makes such a meeting optional. As a matter of custom and practice it is generally accepted that, except in the case of disputes of a relatively small nature, it is best to get the parties together at a preliminary meeting to allow the opportunity for discussion between two people who may not have spoken civilly to one another for some time. It also gives the opportunity for the arbitrator to tell the parties of the likely financial commitment they may be letting themselves in for.

Both sets of rules include provisions seeking to assist in the early development of the right procedure for the arbitration. The parties may, in the case of the ICE Procedure, and must, in the case of CIMAR, make submissions at an early stage aimed at making sure that all parties are aware of the nature of the dispute and so that appropriate procedures can be put into place.

1.9   The powers of the court

The 1996 Act's principles are set out in section 1 which includes, as discussed previously, the right of the parties to agree how their dispute is to be resolved. There has to be a very necessary corollary to this to avoid unnecessary interference with the process by the court. This is set out in section 1(c) which uses the following words:

'in matters governed by this Part the court should not intervene except as provided by this Part.'

The use of the word 'Part' relates to Part I of the 1996 Act (arbitration pursuant to an arbitration agreement) and for most purposes could be mentally replaced by the word Act.

We have advanced from the provisions of the 1950 Act which stated that the authority of the arbitrator was irrevocable except by leave of the High Court and the parties do at least now have the endorsement of the 1996 Act if they wish to make their own decision in this respect. Whether parties were ever effectively precluded from agreeing that an arbitrator whom both of them considered unsuitable should no longer preside over their dispute is in any event a moot point.

There are, however, particular circumstances where the court is needed in the arbitration process, at least in the background. Long gone are the days where arbitral awards were honoured as a matter of course and when failure to do so was a breach of the trust of the remainder of the commercial community which would refuse to trade with the defaulting individual. Nowadays at the very least there is the need to have the support of the court to enforce an arbitrator's award.

There are also a number of other circumstances where the court has powers in respect of the arbitration process. The majority of these are however only there to take over where the parties are unable to reach an agreement.

These are listed in Appendix 1 Table 1(f).

1.10   The mandatory provisions of the 1996 Act

Whilst in general the 1996 Act allows the parties the freedom to agree their procedure and to choose either to be bound by the Act as written or to apply other principles there are certain provisions in the Act that the parties cannot amend even by agreement. These are known as the 'mandatory' provisions and are listed in Schedule 1 to the 1996 Act. The most important of these provisions are those included in sections 33 and 40 which have been discussed above. Other provisions which are mandatory include the power of the court to stay legal proceedings where there is an arbitration agreement, the power of the court to remove an arbitrator, the immunity of the arbitrator and section 66 which covers the enforcement of the award. All the mandatory provisions are matters which if not in place would put the arbitral process into jeopardy. The full list of mandatory provisions is set out in Appendix 1 Table 1(a).

Chapter Two
Before the Hearing

2.1 Introduction

Where the building contract is in standard form the arbitration agreement will specify the manner in which disputes can be referred to arbitration. All the commonly used building standard forms in fact make provision for the party wishing to start the arbitration either to request or to give notice to the other party to agree on a single arbitrator. Failing agreement, the appointment will be made – on the application of either party – by the President or a Vice-President of an appropriate professional body.

This complies with section 16 of the 1996 Act which states that the parties are free to agree on the procedure for appointing the arbitrator. If the unlikely situation arises that there is an arbitration agreement with no procedure for the appointment of the arbitrator or arbitrators, section 16 sets out procedures for the parties to follow. If these procedures fail, and as it is inappropriate for an act of Parliament to name an appointing body, section 18(3)(d) makes the court the final appointer of an arbitrator. There is the possibility in section 18(3)(a) that the court, in exercising its power in this regard, might utilise the services of an appropriate third party appointer rather than carrying out the task itself.

Since arbitration is based on agreement, it is best if the parties are able to agree on the appointment of an arbitrator of their own choice. Indeed, the reason why none of the standard forms provide for the naming of an arbitrator in advance is that he should be chosen in light of the nature of the dispute after it has arisen. It can then be decided whether the dispute should best be referred to an architect, engineer, surveyor, lawyer or builder.

Merely being professionally qualified or having technical knowledge of the subject-matter of the dispute is insufficient in itself. An arbitrator needs other skills, notably the capacity to judge objectively on the basis of evidence and a sound basic knowledge of contract law, the law of tort and arbitration law and practice. If there are difficulties in finding a suitable person, the professional bodies will always be willing to provide a list of suitable people if their advice is sought.

Once an arbitrator is appointed it is in the interests of the parties that he should fulfil his duties as arbitrator and complete the reference either by making an award or awards on all the matters in dispute or by confirming the terms of a settlement between the parties. If the parties have chosen their arbitrator carefully this should be the inevitable outcome unless for some reason the arbitrator is unable to continue. If the appointment has not been wisely made, and an unsuitable arbitrator has been chosen by the parties themselves or a third party appointer, the parties may find themselves in great difficulty. Today, since most of the professional bodies appoint only arbitrators who are known by them to be experienced in arbitration or are listed in the panels of the Chartered Institute of Arbitrators, this disadvantage is no longer so serious as it once was. However, instances have been known, even quite recently, where the parties having met the arbitrator for the first time have then realised his total unsuitability. Should an unsuitable arbitrator be appointed the situation since the 1996 Act is much improved. Previously the only way to obtain the removal of an arbitrator was to seek the leave of the court. Now section 23 of the 1996 Act allows the parties themselves to revoke his authority by agreement. It is not however possible for one party alone to achieve this result in the face of opposition from the other. The only recourse is then to the court and section 24 of the 1996 Act sets out in some detail the grounds which the court will consider.

There are no provisions governing the resignation of an arbitrator in the 1996 Act. It is not considered to be in the interests of the arbitral process that arbitrators should do this and such an action is generally to be deprecated. The parties and the arbitrator may however agree that resignation is the best course. Alternatively the arbitrator himself may reach the conclusion that circumstances have arisen where his resignation will serve the best interests of the resolution of the dispute between the parties.

The 1996 Act does however acknowledge in section 25 that resignation may occur and this section deals with the consequences of such an occurrence. In the case where resignation occurs very early in the process, the involvement of the court is clearly impractical but the question of fees wasted by a party may arise. The arbitrator is of course entitled to recover the cost of any involvement that he may have had, however slight, as a result of the appointment. Practically speaking the overall costs of the arbitration may well be reduced by the arbitrator's resignation but this does not necessarily satisfy a party having to pay out even a relatively small sum. If there is any question as to the responsibility for wasted fees where there has been a change of arbitrator the court draws a parallel with *Queen* v. *Master Manley-Smith* (1893) 63 LJQB 171, which dealt with the consequences of an arbitrator dying. In the unreported case of *James R Morton* v. *Broadway Developments Limited* QBD 1996M 1806 it was held that the costs arising during the appointment of the first arbitrator formed part of the costs of the reference and were to be dealt with by the replacement arbitrator.

Where an arbitrator ceases to hold office there have to be provisions to enable a new arbitrator to be appointed. The 1996 Act covers this in section 27 and, as with many other provisions places the prime responsibility upon the parties to reach agreement as to how the vacancy is to be filled and how much of the previous proceedings should stand.

2.2   *Appointing the arbitrator*

The first step in getting an arbitrator appointed is for the party seeking to start arbitration to write to the other party giving notice of the reference of the dispute to arbitration and requesting him to agree on the appointment of an arbitrator. Figure 2.1 is a typical Notice to Refer which can be used if a party does not have ready access to a list of arbitrators. It is however advisable if possible to give a list of at least three names of potential arbitrators, none of whom should have any subsisting connection with either party or with the subject-matter of the dispute. Before putting forward any names, it is sensible to check informally with those proposed to ensure that they are willing and able to accept the appointment if offered. This may be done by telephone but as the enquirer often wishes to have an early sight of the prospective arbitrator's curriculum vitae and of any conditions that the arbitrator may wish to apply to his appointment, (e.g. fees, payment terms) it is more usually done by post or by fax.

It is important to ensure that the nature of the dispute is properly identified in order to avoid the possibility of disagreement later as to the extent of the matter referred to the arbitrator.

The other party may well reply suggesting other names, but if the parties cannot agree, or the other party does not reply at all, application may be made to the appointing body named in the Appendix to the contract for an appointment. The right to apply to the appointing body depends upon the terms of the contract but

---

**FIGURE 2.1 – NOTICE TO REFER A DISPUTE
TO ARBITRATION**

In the Matter of the Arbitration Act 1996

and

In the Matter of an Arbitration

between

[*Name of Claimant*]   CLAIMANT

and

[*Name of Respondent*]   RESPONDENT

NOTICE OF ARBITRATION

The Claimant hereby gives notice under the Agreement dated [*date*] relating to works executed by the Claimant at [*address*] that it requires the following dispute or difference to be referred to arbitration in accordance with the provisions of clause [*insert arbitration clause number*] of the aforesaid Agreement.

[*Details of the matter[s] to be referred*].

Unless we are able to agree the name of a suitable arbitrator within 14 days of today's date, it is our intention to apply, as provided in the contract, to the President or a Vice-President of [*name of appointing body in contract*] for him to appoint an arbitrator.

Date . . . .

Signed . . . . for and on behalf of the Claimant

---

otherwise will arise after 28 days from the date of the original written notice (1996 Act section 16(3)). The next stage in the appointment process depends upon whether an arbitrator is agreed or is to be appointed in default.

2.2.1 *Where arbitrator agreed*

It is normal for the claimant to write to the arbitrator-designate; alternatively, the parties can write jointly. If only one person writes to him he must, of course, ensure that the other party has in fact agreed to his appointment.

Figure 2.2 is a typical Notice to Concur.

FIGURE 2.2 – NOTICE TO CONCUR IN THE APPOINTMENT OF AN ARBITRATOR

In the Matter of the Arbitration Act 1996

and

In the Matter of an Arbitration

between

[*Name of Claimant*]   CLAIMANT

and

[*Name of Respondent*]   RESPONDENT

NOTICE TO CONCUR IN THE APPOINTMENT OF AN ARBITRATOR

The Claimant hereby gives notice under the Agreement dated [*date*] relating to works executed by the Claimant at [*address*] that it requires the following dispute or difference to be referred to arbitration in accordance with the provisions of clause [*insert arbitration clause number*] of the aforesaid Agreement.

[*Details of the matter[s] to be referred*].

We give below the names of three persons nominated by the Claimant for appointment as arbitrator by agreement:

[*Names and addresses of suggested arbitrators*]

Failing receipt of your agreement to the appointment of one of these persons within the period specified in the contract or your proposal of a name acceptable to us, it is our intention to apply as provided in the contract Agreement, to the President or a Vice-President of [*name of appointing body in contract*] for him to appoint an arbitrator.

Date . . . .

Signed . . . . for and on behalf of the Claimant

The arbitrator-designate must ensure that he has authority to act and so will require to see the arbitration agreement itself or be told what standard form agreement is involved and that the arbitration provision is unamended, so that he may check what the extent of his jurisdiction and powers will be and that he has any special qualifications or knowledge required.

Figure 2.3 is a suitable letter written by the claimant.

FIGURE 2.3 – LETTER TO PROPOSED ARBITRATOR

[*Agreed arbitrator's name and address*]   [*date*]

Dear Sir,

In the matter of the Arbitration Act 1996 and in the matter of an arbitration between [*Name of Claimant*] (Claimant) and [*Name of Respondent*] (Respondent)

A dispute has arisen between us under a contract in the [*version*] Form of Building Contract [*edition*]. We have agreed with the Respondent to appoint you as arbitrator under Clause [*no.*] of that contract, [in the standard form.] [amended as detailed in the enclosed copy.]

We enclose a copy of the Notice to [Refer this Dispute to Arbitration] [Concur in the Appointment of an Arbitrator] which sets out the nature of the dispute between us.

We shall be pleased to hear from you whether you are willing and able to accept this appointment. Would you please let us have details of your terms.

The name and address of the Respondent's representative is [...........].

Yours faithfully

cc [*Respondent*]

The letter or its enclosure may contain this information. If it does not Figure 2.4 provides a specimen letter from the arbitrator-designate. Once aware of these details and assured that he is appropriate for the appointment, the proposed arbitrator should write to the parties setting out his conditions for accepting the appointment, including his terms as to fees, etc. He would be well advised not to accept the appointment without obtaining the parties' agreement on these matters and should defer acceptance until agreement is reached. He should not accept the appointment conditionally upon the parties' agreement to his terms nor should he accept without agreement. If he adopts the latter course he would be entitled to 'reasonable remuneration' only for time spent on the arbitration.

There is no recommended scale of fees for acting as an arbitrator in a UK building arbitration. For example, the *RIBA Architects*

**FIGURE 2.4 – LETTER FROM ARBITRATOR-DESIGNATE
SEEKING FURTHER INFORMATION**

To [*Claimant*]   [*Date*]
c/o [*Claimant's Representative's name and address*]

To [*Respondent*]
c/o [*Respondent's Representative's name and address*]

Dear Sirs,

**In the matter of the Arbitration Act 1996 and in the matter of an
arbitration between [*Name of Claimant*] (Claimant) and [*Name of
Respondent*] (Respondent)**

I thank the Claimant for its letter of [*date*] and for the Notice to
Concur enclosed therewith. Before I respond I should like to satisfy
myself as to the terms of the contract between you and that there is a
binding arbitration agreement.

Would the Claimant please send me a copy of the actual contract as
soon as possible.

Yours faithfully

Appointment (1982), part 2.36 envisages the architect acting as arbitrator, but makes no recommendation on the scale of payment for
these services. It is usual to charge on the basis of time involved, and
part 4.9 provides:

> 'Time charges are based on hourly rates for principals and other
> technical staff. In assessing the hourly rate, all relevant factors
> should be considered, including the complexity of the work, the
> qualifications, experience and responsibility of the architect, and
> the character of any negotiations. Hourly rates for principals shall
> be agreed ...'

By way of guidance, at the time of writing, in most cases fees
charged by technical arbitrators will fall somewhere between £600
and £1200 a day for hearings, etc. and £80 to £175 an hour for other
time. VAT is added if appropriate. Much will depend upon the
qualifications and experience of the arbitrator, the nature of the case,
etc. The actual rate charged will depend upon the individual arbitrator, based upon an analysis of his office and other overhead costs.

The important matters to be settled are:
- Daily and hourly rates for time spent.
- Cancellation and postponement charges.
- Reimbursement of expenses.
- Security for the arbitrator's fees
- Effect of inflation.
- Period for taking up the award.
- Procedural rules, if any.

Figure 2.5 is a typical letter from the arbitrator-designate agreeing to act but before doing so seeking agreement to his various terms. Figure 2.5A is the Form of Appointment referred to in Figure 2.5.

FIGURE 2.5 – LETTER FROM ARBITRATOR-DESIGNATE

To [*Claimant*]   [*Date*]
c/o [*Claimant's Representative's name and address*]

To [*Respondent*]
c/o [*Respondent's Representative's name and address*]

Dear Sirs,

In the matter of the Arbitration Act 1996 and in the matter of an arbitration between [*Name of Claimant*] (Claimant) and [*Name of Respondent*] (Respondent)

Further to the previous correspondence between us, I confirm that I am willing and able to accept appointment as arbitrator in the dispute between you. My acceptance is, of course, subject to acceptance of my terms for acting as arbitrator.

I enclose my proposed Form of Appointment for your consideration which sets out my terms. I should be obliged if the Claimant would sign the Form where indicated and forward it to the Respondent for counter-signature. Once I have received the Form signed by both of you I shall sign it myself and send you each a copy.

Yours faithfully

FIGURE 2.5A – FORM OF APPOINTMENT
OF ARBITRATOR PARTY AGREED APPOINTMENT

The following form is that published by the Society of Construction Arbitrators

FORM OF AGREEMENT FOR THE APPOINTMENT OF AN ARBITRATOR

By the terms of an Agreement in writing dated [*date*]

between [*Name of Claimant*] of [*Address*]

and [*Name of Respondent*]   of [*Address*]

relating to [*Name of Project/dispute*]

it is provided that any dispute or difference arising thereunder shall be referred to arbitration.

A dispute or difference having arisen between the parties to the said Agreement,

the said [*Name of Claimant*] and [*Name of Respondent*]

(hereinafter referred to as 'the parties'), pursuant to the provisions of the said Agreement hereby appoint

[*Name of Arbitrator*]   of [*Address*]

(hereinafter referred to as 'the arbitrator') to be sole arbitrator in this matter upon the following Terms and Conditions.

Fees

1.   Notwithstanding any order as to costs in any award, the parties shall be jointly and severally liable to the arbitrator for the due and timely payment of his fees, costs and expenses in accordance with these Terms and Conditions.

2.   Whether or not a hearing is conducted or an award made, the arbitrator shall be entitled to payments at the rates set out herein for all time spent on or connection with the arbitration and for all time allocated thereto, together with all out-of-pocket expenses and other disbursements reasonably incurred. Where applicable, value added tax shall be charged and paid for in addition to such fees, expenses and disbursements.

3. The arbitrator's fees shall include the following:

    (a) Either (i) an **appointment fee** of £ [   ]
              or (ii) a **minimum fee** of £ [   ]
         which shall be paid to and retained by the arbitrator in any event.

    (b) An **hourly rate** of £ [   ] for each hour or part of an hour spent on or in connection with the arbitration, including but not limited to time spent on general administration, travelling and reading papers.

         Provided that, if no daily rate is specified in paragraph 3(c)(i) below, all time spent on a day set down for hearing the arbitration (including time spent in adjournments) shall rank for payment at the hourly rate, with a minimum of eight hours' payment for that day in any event.

    (c) (i)   A **daily rate** of £ [   ] for each day or part of a day spent in hearings, meetings, inspections, site visits and the like. A day shall be deemed to comprise up to but no more than eight hours including adjournments. Time spent in excess of this limit shall be charged extra at the hourly rate, as shall time spent in travelling and reading papers on a hearing day.
         (ii) Where appropriate, an **alternative daily rate** of £ [   ] for each day not occupied in arbitration business which is necessarily spent away from the arbitrator's normal residence or place of business. Where a day is charged for on this basis, any arbitration work actually undertaken on that day shall be deemed to be covered thereby.

    (d) A **booking fee** of £ [   ] for each day reserved for a hearing, payable by the party or parties requesting the booking and when the booking is made. The arbitrator shall give credit for any booking fees he has received against any other fees which may subsequently become due in respect of the days so reserved.

    (e) A **cancellation fee** for each day reserved for a hearing which is later vacated, whether by adjournment or by cancellation, according to the following scale:
         (i)   Where the booking is vacated more than six months before the first day so reserved:
               A fee equal to [   ] per cent of the daily rate for each day vacated.
         (ii) Where the booking is vacated six months or less but more than three months before the first day so revealed:
               A fee equal to [   ] per cent of the daily rate for each day vacated.

31

(iii) Where the booking is vacated three months or less but more than one month before the first day so reserved:
A fee equal to [   ] per cent of the daily rate for each day vacated.

(iv) Where the booking is vacated one month or less before the first day so reserved:
A fee equal to [   ] per cent of the daily rate for each day vacated.

For the purpose of this paragraph the daily rate shall be that specified in paragraph 3(c)(i) above or, if none is there specified, a rate equivalent to eight times the hourly rate specified in paragraph 3(b) above.

The above scale shall apply whatever the length of the booking vacated.

Payment

4.   Notwithstanding any provision for a payment by way of security, the arbitrator shall be entitled to submit, and the parties shall pay, such interim accounts of fees, expenses and disbursements as the arbitrator thinks appropriate.

5.   Where the arbitration continues for more than one year after the arbitrator was appointed, the arbitrator shall be entitled to re-value his fees in respect of subsequent years generally in line with the General Index of Retail Prices (RPI).

6.   Payment in full of the arbitrator's fees, expenses and disbursements shall be made by the parties within ten days of the date of publication of each and every award, or within forty days of the date of submission of each interim account in accordance with paragraph 4, above. Should the parties fail to make payment in full within the said ten or forty days (as the case may be), interest upon all monies then outstanding shall accrue at a rate of [   ] per cent above the base rate charged from time to time by the arbitrator's Bankers.

General

7.   The arbitrator shall be entitled to require any or all of the parties to pay such amount as he shall determine from time to time to be held by the arbitrator as security for the due payment of his fees, expenses and disbursements.

8.   The arbitrator's disbursements may include the cost of obtaining such legal advice or technical assistance as in his absolute discretion he shall deem it desirable to take.
   Provided that the advice of leading counsel shall not be sought without the express agreement of all the parties to the arbitration.

9. Should a settlement be reached between all or any two of the parties, and whether such settlement disposes of all or only some of the matters at issue, the parties shall forthwith inform the arbitrator of the terms of the settlement so that the same may be incorporated in an award.

Signed and delivered by:

_____   _____

[on behalf of claimant]   *[on behalf of Respondent]*

[date]   *[date]*

This Form of Agreement having been signed by or on behalf of all the said parties, I the said [*Name of Arbitrator*] hereby accept the said appointment upon the Terms and Conditions hereinbefore stated.

_____

[date]   © S. C. A.
                                                    July, 1987

Note: This Form of Agreement was prepared before the 1996 Arbitration Act came into force. It is useful to clarify the Arbitrator's powers under that Act by adding a provision to the following effect:

The parties shall notify the arbitrator of any agreement that they reach which might limit or define his authority or the powers given to him as Arbitrator under the Arbitration Act 1996

Figure 2.5B sets out the explanatory notes published by the Society of Construction Arbitrators to accompany their Form of Agreement.

FIGURE 2.5B – EXPLANATORY NOTES TO FORM OF AGREEMENT
EXPLANATORY NOTES

Introduction

The Society has produced two alternative versions of the Terms and Conditions under which it recommends members to accept appointments in arbitrations arising in the construction industry. The first is in the form of an Agreement to be used when the appointment is made direct by the parties; it would be completed by the parties, signed by them or on their behalf and sent to the selected arbitrator for countersignature to signify his acceptance of the appointment. The second version is for use where the appointment is made by an outside body and where the arbitrator-elect wishes to record the

Terms and Conditions on which he would be willing to accept that appointment. Both versions contain identical Terms and the following Notes apply equally to either.

These Explanatory Notes do not form part of the Agreement and are issued by the Society for information and guidance only.

GENERAL

1. Paragraph 1 of the Terms emphasises that the parties are jointly and severally liable for the payment of the arbitrator's fees and expenses. If, for any reason, a party held liable to pay the arbitrator a sum of money in respect of fees and/or expenses fails to do so within a reasonable time, this provision enables the arbitrator to recover the money from the other party or parties.

2. The arbitrator's right to be paid is unconditional. Paragraph 2 of the Terms makes it clear that this right is *not* dependent upon his holding a hearing or publishing an award, and applies even if the dispute is settled by amicable arrangement between the parties during the course of the reference.

3. While the Society recognises that there are several other bases upon which an arbitrator's fees can be calculated, it considers that payment based on time occupied in the duties of the reference is the fairest, both to the arbitrator and to the parties. Paragraph 2 also stresses that the basis for fees is the *total* time spent on or in connection with the arbitration and the attention of the parties is drawn to the fact that an arbitrator is entitled to charge for time allocated to the reference but which, for some reason outside the control of the arbitrator, is not so used.

4. The arbitrator's **total fee** is made up of either three or four unit charges selected from items set out in paragraph 3 of the Terms, namely:

    (i) *either* 3(a)(i) *or* 3 (a)(ii); and
    (ii) *either* 3(b) *or* a combination of 3(b) and 3(c)(i) and maybe 3(c)(ii); and
    (iii) 3(d); and maybe
    (iv) 3(e).

    It should be noted that if alternative 3(a)(i) is adopted then the appointment fee is retained by the arbitrator in any event; it is part of his total fee and does not have to be credited against any later payment. If, however, alternative 3(a)(ii) is adopted and the minimum fee is paid to the arbitrator on appointment, then the sum so paid must be credited against any larger fee which may later become payable under other provisions of paragraph 3 of the Terms.

5. The **appointment fee** (pargraph 3(a)(i)) is intended to cover the necessary administrative work involved in correspondence with the parties or the appointing body, the preparation and completion of the Form of Appointment, the initial letter of acceptance and the opening of a new file and associated records.

6. The **minimum fee** (paragraph 3(a)(ii)) should include not only for the work described in the preceding paragraph but also for some measure of preliminary work and/or involvement or inconvenience for which the arbitrator anticipates that he would not be adequately compensated by a time charge should the reference be settled or abandoned at an early stage.

7. The **hourly rate** (paragraph 3(b)) is intended to be the normal basis for the calculation of fees. When a day has been set aside for a hearing, all the time occupied on that day for whatever purpose is chargeable, with a minimum of 8 hours. It is, however, up to each arbitrator to decide for himself whether or not to include a charge for regular daily adjournments such as lunch and tea breaks. In reaching this decision regard will doubtless be given to whether or not the arbitrator customarily works on the arbitration during such adjournments or has the opportunity of earning fees by other work.

For example, a day set aside for a hearing might result in the following timetable:

0845 to 0915	Travel to hearing venue.
0915 to 0945	Preparing papers for the hearing.
0945 to 1000	Waiting to start the hearing.
1000 to 1300	Hearing.
1300 to 1400	Lunch adjournment.
1400 to 1430	Adjournment at Respondent's request.
1430 to 1545	Hearing.
1545 to 1600	Packing up papers etc.
1600 to 1630	Travel back to office.

The timetable set out above covers a total elapsed of $7\frac{3}{4}$ hours (or $6\frac{3}{4}$ hours if the lunch adjournment is not counted), so the minimum would apply and the day would be charged at 8 hours.

However, if the day's timetable was as follows:-

0800 to 0915	Travel to hearing venue.
0915 to 0945	Reading papers.
0945 to 1000	Waiting to start the hearing.
1000 to 1300	Hearing.
1300 to 1430	Lunch adjournment.
1430 to 1730	Hearing.
1730 to 1745	Packing up papers etc.
1745 to 1900	Travel back to office.

the total elapsed time would be 11 hours (or $9\frac{1}{2}$ hours if the lunch adjournment is not counted) and, as the time occupied exceeds the minimum, the day would be charged in full as 11 hours or $9\frac{1}{2}$ hours as the case may be.

While not essential, most arbitrators will find it desirable to keep records of the time actually occupied on the duties of the reference and, if the daily rates (paragraphs 3(c)(i) or 3(c)(ii)) are adopted, such records should be separated into the different time categories of reading papers, travelling and the various duties of the reference to facilitate the calculation of proper fees.

8.   The **daily rate** (paragraph 3(c)(i)) Some arbitrators prefer to base their charges for hearings, inspections, meetings and site visits etc. on a daily rather than an hourly rate. While provision for this is made in the Terms, it should be noted that it will apply in conjunction with the hourly rate which will still be used for any time in excess of 8 hours in a day and for all travelling time. Assuming that the arbitrator decides to charge for lunch adjournments, the timetables discussed in the preceding paragraph would result in:

> for the first example, a charge of one day at the daily rate plus $\frac{3}{4}$ hour for travelling at the hourly rate; and

> for the second example, a charge of one day at the daily rate plus $2\frac{1}{2}$ hours for travelling at the hourly rate and a further $\frac{1}{2}$ hour at the hourly rate for the excess time over the 8 hour minimum.

9.   The **alternative daily rate** (paragraph 3(c)(ii)) will normally be applicable only for arbitrations where the hearing, site inspections or meetings take place outside the United Kingdom. It is intended to cover weekend and other short adjournments (including local public holidays) where it is not practicable for the arbitrator to return home and is thus precluded from handling other work.

10.   The **booking fee** (paragraph 3(d)) The purpose of the booking fee is to stress to the parties that by formally reserving time for a hearing the arbitrator is precluded from accepting any other fee-earning activity in the same period. The requirement that the parties should pay a booking fee when asking the arbitrator to reserve time for a hearing takes cognisance of this fact and helps to ensure that the parties are neither over-optimistic in setting early hearing dates nor over-generous in estimating the likely duration of the hearing. If the arbitration proceeds to a hearing for which full fees are ultimately chargeable, of if a cancellation fee becomes payable under paragraph 3(e) of the Terms, then the booking fee will be credited in full.

11. **Cancellation fees** (paragraphs 3(e)(i) to (iv)) The Society considered many different methods by which an arbitrator should be able to receive payment in respect of time set aside by him for the duties of the reference but not actually so used for reasons outside his control. It was accepted that the nearer to the starting date of a hearing the cancellation occurred, the greater should be the sum payable to the arbitrator.

    The graduation of payments by calendar months is the most direct approach, and each arbitrator is free to insert the percentages which are appropriate to his practice and circumstances. It should be realised, therefore, that the percentages inserted by one arbitrator will be no guide to these adopted by another arbitrator.

    Cancellation fees apply irrespective of the length of the hearing.

12. **Security payments** (paragraphs 4 and 7) Security payments differ from payments on account in that the former are received before the fees are earned and may have to be accounted for at a later date (and possibly returned in part), whereas the latter are received in respect of fees actually earned and are not returnable even though such payments may be brought to account when rendering the final fee account.

13. The provision in paragraph 5 of the Terms for the revaluation of fees where the reference continues for more than a year is essential in periods of high inflation and desirable in less inflationary times because of the long period which can elapse between appointment and award in construction cases. It is important to note the words '... generally in line with the General Index of Retail Prices...'. It is not intended that slavish regard should be had to the RPI, which would result in fees being charged at rates involving fractions of a penny, but rather that a comparison of the RPI at the date of appointment with the RPI a year later should indicate a figure which, after rounding up or down as appropriate, would produce a reasonable increase in fee.

14. While most arbitrators receive their fees promptly on publication of their awards, there are exceptions and it is considered that, where payment is unreasonably delayed, the arbitrator should be entitled to levy interest on the outstanding sum. This is in line with current commercial thinking. The period of 10 days after publication of an award is also in line with the provisions of R.S.C. Order 73 which allows the parties only 21 days from the date of publication in which to lodge any Motion for leave to appeal, while 40 days' grace after the submission of an interim allows ample time for the slowest of administrations to

effect payment. The blank in paragraph 6 allows the individual arbitrator to decide on the level of interest he would require.

15. Paragraph 9 of the Terms confirms the usual practice that the terms of a negotiated settlement should be confirmed in an award so that the settlement can, if necessary, be enforced or to enable costs to be taxed if they are not agreed between the parties.

THE FORM OF AGREEMENT

16. The date, the names of the parties and the definition of the 'Works' to be inserted in the opening sentence of the recitals should be copied from the Head Contract which includes the arbitration clause.

17. In the recitals provision is also made for the insertion of the name and address of a third party in order to cover disputes involving (for example) a sub-contractor where the three parties have agreed that the hearings should be consolidated. In such a case it may be necessary in the opening sentence to refer to two separate agreements.

COPYRIGHT

18. The copyright in these Explanatory Notes and in the Form of Agreement rests with the Society. However, Members of the Society may have the Form of Agreement printed up to incorporate their own names and addresses provided that the personalised Form so produced contains due acknowledgement of the Society's copyright.

© S. C. A.
July, 1987

Once these matters are settled, the arbitrator signs the form in acceptance of the appointment and sends copies to each party. He then convenes a preliminary meeting or issues directions should a preliminary meeting not be considered appropriate.

Figure 2.6 completes the appointment.

FIGURE 2.6 – LETTER COMPLETING APPOINTMENT OF ARBITRATOR AND ARRANGING A PRELIMINARY MEETING

To [*Claimant*]   [*Date*]
c/o [*Claimant's Representative's name and address*]

To [*Respondent*]
c/o [*Respondent's Representative's name and address*]

Dear Sirs,

In the matter of the Arbitration Act 1996 and in the matter of an arbitration between [*Name of Claimant*] (Claimant) and [*Name of Respondent*] (Respondent)

I confirm receipt of my Form of Appointment from the Respondent signed on behalf of both parties and I thank you for agreeing to my terms. I have today signed the Form myself in acceptance of the appointment and I enclose a copy for your records herewith.

I propose to hold a Preliminary Meeting in order to consider the procedures and programme and to discuss the Directions I need to issue for the future conduct of this arbitration. I offer the following dates for a Preliminary Meeting which I have available at this time; [*two or three dates*] (morning only in each case). I suggest that the meeting starts at 10.30AM. The venue should be agreed between yourselves but is otherwise immaterial to me. Would the Claimant please make the arrangements for the meeting, agree them with the Respondent and inform me when that is done.

I enclose my standard agenda and checklist for the meeting for your attention. This may include matters that it is too early to discuss at this time and these matters can be left in abeyance if this is considered appropriate.

Yours faithfully

Note: Some arbitrators like to formalise their letters to the parties by endorsing their name at the foot with the word 'Arbitrator'. All letters from an arbitrator should be on headed paper in his own name with his qualifications and the fact that he is arbitrator set out thereon. (A company headed notepaper should not be used.)

2.2.2 Where arbitrator appointed by third party

Where the reaction of the employer/contractor to the notice to concur (Figure 2.2) is one of refusal of all the named prospective arbitrators, the reply often includes a further list of three suggested names. The applicant's reaction to this is quite often one of 'if they won't accept my list, I won't accept theirs'. This is a somewhat unfortunate reaction as the end result is usually the appointment of an arbitrator by the appointing body named in the contract. As discussed earlier this is now rather less of a lottery than in former times due to the general implementation of training procedures and quality control of lists of arbitrators. It does however mean that the parties lose control of who is appointed as their arbitrator and they *may* get someone who is unsuitable.

The applicant should, upon it becoming evident that an arbitrator will not be agreed and after expiry of any time set in the contract for that agreement, obtain the necessary application forms from the body named in the contract as the appointing authority. All appointing authorities have their own standard forms which have minor differences but the principal aim of all of them is to identify the existence of an arbitration agreement and to seek sufficient information concerning the nature of the dispute in an endeavour to appoint the most appropriate arbitrator.

One common objection by an employer/contractor to an application by a contractor or sub-contractor for the appointment of an arbitrator by an appointing authority is that the requirements of the arbitration agreement have not been followed properly. Often the objection is that there has been no attempt to agree an arbitrator before the application was made. This was the situation that arose in *Essoldo Limited* v. *Elcresta Limited* (1972) 23 P & CR 1; (1971) 220 EG 1437, a rent review matter, and it was held that there was no justification in construing the phrase 'in default of agreement' as if there were added the words 'after an attempt has been made...' and that a party could refer the matter to the appointing authority notwithstanding that it had made no attempt at all to agree. This means that unless the terms of the contract actually require the issue of a notice to concur in the appointment of an arbitrator, the failure to issue such a notice does not preclude an application to the appointing authority named in the contract after the requisite period for agreement has elapsed. This does of course mean that the parties then lose the opportunity to agree the most suitable arbitrator with the resultant disadvantages already mentioned.

In this connection it is interesting to compare the difference between the versions of clause 41.1 in the JCT contracts pre and post 1988 when the requirement for a notice to concur as a condition precedent for an application to an appointing authority was removed and replaced by a simple requirement to give notice of arbitration. This means that the situation as defined in *Essoldo* has applied to the JCT contracts since 1988.

Another situation that sometimes arises is that an objection is made to the appointing body by the non-applicant that any appointment if made would be invalid or that there is no authority for the appointment to be made. In *United Co-operatives* v. *Sun Alliance & London Assurance Co Ltd* (1987) 1 EGLR 126 an injunction was sought to restrain the President of the RICS from making an appointment on the grounds that the application was premature and thus invalid. The court looked into the policy of the President. Its conclusion was that when asked to appoint he should do so.

Once the application is received the appointing body will ascertain the name of a prospective appointee and check with him a number of matters. These will include points such as whether he has any connection with the parties and that he is available and willing to act. It will then issue a form of appointment signed by the President or a Vice-President naming the arbitrator and send copies to each party and to the arbitrator himself.

The question sometimes arises as to whether the appointment is complete upon the issue of the form of appointment. The view taken by most appointing authorities is that this is so. The RIBA has an alternative view allowing the arbitrator the opportunity to agree his terms and conditions with the parties before he returns the form of appointment endorsed with his signature confirming his acceptance of the appointment.

Upon receipt of the institutional appointment the arbitrator will need to inform the parties of the terms that he wishes to apply to his appointment. There are different views as to the appropriate procedure here. Some arbitrators seek the parties' agreement to their terms even where the appointment is by an institution that views the appointment as being complete. The arbitrator will, in this instance, change his form of appointment into terms and conditions and send it to the parties. In many cases his proposals will be accepted by both parties but it is sometimes the case that the one or both parties take several objections to the terms, or even refuse to agree to them at all.

In *K/S Norjarl A/S* v. *Hyundai Heavy Industries Co Ltd* [1992] 1 QB 863; [1991] 3 All ER 211; [1991] 3 WLR 1025, CA it was held that it probably amounted to misconduct for an arbitrator to negotiate and agree terms with one party alone after he had accepted the appointment since this would make the arbitrator vulnerable to an imputation of bias.

There is nothing however to prevent an arbitrator who is in receipt of an institutional appointment from negotiating his terms with the parties provided that he does it with both of them and it is sometimes the case that a revised form of agreement results and the arbitrator proceeds on that basis. Where the objection is substantial it is inappropriate for the arbitrator to refuse to act given the source of his appointment and the normal procedure is for the arbitrator to tell the parties that he intends to proceed and will rely upon his right to reasonable fees and expenses as are appropriate in the circumstances ((section 28(1) of the Act).

Other arbitrators will seek to pre-empt this occurrence and the delays that can occur and will merely send their terms and conditions to the parties stating the terms that they will apply to the appointment and rely upon their right to a reasonable fee. The need for any possible retraction in the case of the possibly difficult situation where the claimant agrees and the respondent refuses to agree anything at all is of course avoided where this procedure is utilised.

The possibility of the resignation of the arbitrator has been mentioned earlier. One instance in which this may occur is where the arbitrator is appointed by an appointing body on the basis of an application of one of the parties. He checks that he has no connection with either party and replies to the appointing body to that effect. He then, once appointed, finds out that one party is represented by a solicitor who has appointed him in the past as expert witness. This is fine if the other party takes no objection and he must as a matter of course tell the other party in these circumstances. Where, however, there is an objection and this is often the case where objecting party is a respondent who is reluctant to see the arbitration progress, there is the distinct likelihood that accusations of bias will disrupt the smooth progress of the arbitration when, as is almost inevitable, the arbitrator decides something against the party making the objection. The arbitrator can, of course, soldier on and may well decide to do so.

There could be circumstances, however, where the arbitrator considers that his withdrawal will be in the best interests of the smooth running of the arbitration and the resolution of the dispute. Unless both parties agree to this effect, it is for the arbitrator to make the final decision on the basis that he should avoid unnecessary delay and expense, section 33(1)(b). The appointment procedure then starts again, the wasted costs forming part of the costs of the reference (see above).

Figure 2.7 is a letter from an arbitrator appointed by an institution to the parties.

FIGURE 2.7 – LETTER TO PARTIES ON RECEIPT OF INSTITUTIONAL APPOINTMENT AND ARRANGING PRELIMINARY MEETING

To [*Claimant*]   [*Date*]
c/o [*Claimant's Representative's name and address*]

To [*Respondent*]
c/o [*Respondent Representative's name and address*]

Dear Sirs,

In the matter of the Arbitration Act 1996 and in the matter of an arbitration between [*Name of Claimant*] (Claimant) and [*Name of Respondent*] (Respondent)

I am in receipt of an appointment as arbitrator by the President of [*name of institution*]. I have accepted this appointment.

I enclose my Terms and Conditions for acting as arbitrator [*which I invite you to agree*].

I propose to hold a Preliminary Meeting in order to consider the procedures and programme and to discuss the Directions I need to issue for the future conduct of this arbitration. I offer the following dates for a Preliminary Meeting which I have available at this time; [*two or three dates*] (morning only in each case). I suggest that the meeting starts at 10.30AM. The venue should be agreed between yourselves but is otherwise immaterial to me. Would the Claimant please make the arrangements for the meeting, agree them with the Respondent and inform me when that is done.

I enclose my standard agenda and checklist for the meeting for your attention. This may include matters that it is too early to discuss at this time and these matters can be left in abeyance if this is considered appropriate.

Yours faithfully

Note: The Arbitrator's terms and conditions in this instance would be similar to those in Figure 2.5A above save that amendments will have to be made to cater for the appointment method and the provision for party signatures may be omitted as discussed above.

2.3   *The preliminary meeting*

It is usual in construction arbitrations for the arbitrator to start the arbitration process by calling a preliminary meeting so that he may discuss and agree with the parties the way in which the arbitration shall proceed. Sometimes, if the dispute is a small one and the parties are geographically far apart from each other and the arbitrator, a preliminary meeting may be dispensed with and the matters dealt with by correspondence. This should be the exception, since it is always wise for the arbitrator to meet the parties at an early stage so that he may get to know them and they can get to know him. It is also often possible at this stage for the arbitrator to get a feel of the dispute and he may be able, at the meeting, to suggest a means of resolving the matter without resort to a full arbitration procedure.

Suppose, for example, there is a very small amount of money at stake, the arbitrator might well at the preliminary meeting point out to the parties that the eventual costs involved will substantially exceed the amount in dispute, a point which in the heat of their argument they might not have appreciated. If they are legally represented their advisers should, of course, have already done this, but a substantial number of disputes are still compromised at or shortly after the preliminary meeting.

In the case of an appointment agreed with the parties the arbitrator will have already set up a procedure for arranging the preliminary meeting by his letter as Figure 2.6. With an institutional appointment where the appointment is finalised on signature of the form by the President, the arbitrator should take the initiative and get the reference moving by offering alternative dates for a preliminary meeting at the same time as sending his terms to the parties unless there is some good reason for not doing so.

When the time comes for the meeting the arbitrator should ensure that both parties and/or their representatives have arrived before he enters the room or admits them to his room. He should have with him:

- An agenda and questionnaire.
- A notebook in case matters arise which are not covered by the questionnaire or checklist but which should be recorded.
- The file containing the details of his appointment and any documents he has so far received.
- Any reference works on procedural and allied matters.
- A copy of the procedural rules agreed or proposed.

Figure 2.8 is an example of an agenda and questionnaire.

FIGURE 2.8 – AGENDA AND QUESTIONNAIRE FOR PRELIMINARY MEETING

1. Preliminary meeting: Date   Venue

2. Attendance: For Claimant
                For Respondent

3. Confirmation of arbitration agreement

4. Confirmation of names of Parties
   Any changes since contract?

5. Names of representatives and addresses
   for communications: For Claimant
                        For Respondent

6. Confirmation that Arbitrator is properly
   appointed
   Terms and Conditions agreed?

7. Arbitration Act 1996
   Any agreements to limit or define
   Arbitrator's authority and powers from
   those set out in the Act?
   Seat of Arbitration?

8. Brief preliminary description of dispute
   Amount in dispute: Claim
                       Counterclaim

9. Extent of Arbitrator's jurisdiction
   To cover all matters in dispute arising
   out of or under or consequent upon the
   contract?
   or
   Limited by reference to arbitration?

10. General consideration of procedures
    Cost considerations (see section 33)

11. Will taking preliminary issue(s)
    encourage early resolution of dispute?

12. Arbitration Rules applicable to reference
    under contract
    Are any Arbitration Rules to be adopted?

13. Documents only procedure or hearing?

---

14. Pleadings or full Statements of Case?

15. Timetable including date deemed to be
     'Close of Pleadings'

16. Chronology of events

17. Service of Pleadings/Statements by fax
        If this is done it should be to the other
        Party only
        To Arbitrator by post (covering letter
        only copied by fax to Arbitrator if
        party wishes to confirm that service
        has been achieved within time set)

18. Software capabilities if service or
     exchange by electronic means?

19. Scott schedules

20. Requests for Further and Better
     particulars
        By leave on application to arbitrator

21. Documents submitted with Statements
     to be deemed accepted by other Party
     and thus admissible evidence in the
     arbitration unless objected to in writing
     at the time of or before service of the
     following Statement

22. Preliminary list of issues to be agreed
     within 14 days of 'Close of Pleadings'.

23. Discovery (see section 34(2)(d))
        If ordered to be limited to preliminary
        list of issues
        14 days after 'Close of Pleadings'
        Provision of copies of documents
        (reasonable copying costs to be paid)

24. View of site
        Before/during/after hearing?

25. Total number of witnesses to be limited
     in the interests of overall cost?

26. Witnesses of fact

    Date for exchange of Statements

    Statements to be evidence in chief
    The Parties are recommended in the
    interests of economy to limit witness
    statements to principal matters only.
    Oral examination in chief to cover
    matters excluded from a statement
    will, if this is done, not be prohibited.

    Form of Statement – Affidavit if
    'documents only' procedure?

27. Expert witnesses

    Limitation of experts of the Arbitrator's
    own discipline in the interests of cost?
    Use by Arbitrator of own expertise?
    Arbitrator to meet in-house Quantity
    Surveyors to resolve quantum?

    Number on each side

    Experts to meet to narrow issues and
    agree facts/figures

    Date for joint open report

    Date for exchange of final reports

    Failure to exchange to debar witness

    Reports to be evidence in chief

28. Any legal submission to be in writing
    Procedural applications may be taken
    orally at the Arbitrator's discretion

29. Further meetings/pre hearing review?
    Use of telephone conference facilities?

30. Bundle:

    One agreed bundle only, in
    chronological order starting from the
    front, paginated and indexed

    Core bundle if necessary
    Numbering as main bundle

    Drawings, documents, figures,
    photographs, etc. to be agreed as far as
    possible.

    Photographs to be numbered on face.

31. Procedure prior to hearing:

> Written opening addresses and core bundle to be submitted to the Arbitrator one week before start of hearing

> Hearing bundle to Arbitrator at start of hearing

> Copies of law reports and authorities to be supplied to Arbitrator

32. Final list of matters for Arbitrator's decision in written opening

33. Hearing:

> Date and length (total number of witnesses)

> Firm or provisional?

> Venue – Arrangements made by Claimant in consultation with Respondent

> 4 day sitting

> Representation: Claimant
>   Respondent

> Shorthand note/tape recording

> Evidence is to be taken on oath or affirmation?
> Where an oath is to be taken other than on the Holy Bible, the Party calling that witness shall provide the necessary Book(s) and/or other materials and if a language other than English is to be used a translation is to be supplied.

> Strict Rules of Evidence not to apply (see section 34(2)(f))

> Closing addresses in writing
> Time and procedure for submission?

34. Form of award

> Final – submissions on costs at end of hearing or in documentary submissions.
> *or*

Interim – later submissions on costs

A reasoned award including the Arbitrator's reasoning with regard to findings of fact will be given unless it is agreed by both parties that this is not required (see section 52(4))

35. Settlements ratified by Agreed Award or Termination Order if Parties so require.

36. Determining the recoverable costs (see section 63 and 63(1), (2), (3) and (5) in particular)

37. Limiting recoverable costs (see section 65)

38. Communications to Arbitrator always to be copied to other side and so endorsed

39. Costs of meeting and subsequent order to be costs in the arbitration

40. Liberty to apply

41. Any other matters

The arbitrator should not go into the meeting – which is often held at his own office or other neutral ground – unprepared. He should have clearly in mind the course which he thinks the arbitration should take at this stage and the questions which he wants to put to the parties. The provisions of section 1(6) of the Act should however always be borne in mind and it is ultimately for the parties to decide, albeit with his advice, how the arbitration should be conducted.

Most arbitrators have a standard questionnaire or checklist of the points to be covered at the meeting (see Figure 2.8), and, if the meeting is dispensed with, this can be used as the basis of agreement by correspondence. The arbitrator should take the initiative at the preliminary meeting. Particularly if the parties are not legally represented he should ensure that discussion proceeds along the right lines, that both parties have a chance to express their views and that, at the end of the meeting, they are both satisfied and understand what has been decided and what they have agreed to do.

Ideally, all decisions taken at the preliminary meeting should be with the consent of both parties; this is generally possible if they are legally or otherwise professionally represented. The meeting ought to be conducted on formal lines with the arbitrator acting as a firm chairman to ensure that discussion is kept under control and that the parties behave reasonably. Doing this requires a good deal of tact but the arbitrator should never be afraid of taking a firm line.

No two preliminary meetings are the same, but at every such meeting certain points must be established:

- The broad terms of the dispute, the details of which are to be set out in the statement of case or pleadings.
- The way in which the arbitration is to be conducted (e.g. documents only, documents with inspection, or a procedure whereby the issues for the decision of the arbitrator are identified by traditional pleadings or by statements of case either of which being followed by a hearing).
- What procedural rules, if any, will apply.
- The timetable, although at this stage it is probably advisable to fix only provisional dates for a hearing, since much can happen during the intervening stages.
- Whether or not the parties intend to be represented by solicitors or other professionals.

It is also vital that the arbitrator seeks to ascertain the extent of the matters he has jurisdiction to deal with at the meeting. It is not really in the interests of the resolution of a dispute if an arbitrator deals with a matter that the parties have not both agreed to him dealing with, as there is no simple way to remedy an award that deals with something that it shouldn't. There is nothing in section 57 of the 1996 Act, which deals with the correction of awards, to allow correction in such an event. The only formal recourse in the 1996 Act for this situation is to apply to the court under section 67 which is likely to take some considerable time and cost the parties a lot of money. Pragmatic parties could of course agree that the arbitrator should revise his award in these circumstances but as one party may be quite pleased that the arbitrator has overstepped the mark, this is unlikely. (Wider issues of jurisdiction are dealt with in Chapter 3.)

The reverse case of the arbitrator failing to deal with an issue that he was asked to deal with is easier. Section 57 of the Act provides for the preparation of an additional award should the arbitrator fail to deal with a claim that was presented to him but was not dealt with in the award. This correction has to be done within a very limited time period.

The upshot of the above is that the extent of the arbitrator's jurisdiction must be a matter to be confirmed at the preliminary meeting and it is in this respect vital to ensure that the notice of dispute in particular is examined to clarify this point.

2.4 *The order*

After the preliminary meeting the arbitrator must embody the decisions reached and directions he wishes to make in a formal document called an 'Order', a typical example of which is shown as *Figure 2.9*

FIGURE 2.9 – TYPICAL ORDER AFTER PRELIMINARY MEETING

In the Matter of the Arbitration Act 1996

and

In the Matter of an Arbitration

between

[*Name of Claimant*]   CLAIMANT

and

[*Name of Respondent*]   RESPONDENT

ORDER No. 1

Having heard Mr [*********] of counsel on behalf of the Claimant and Mr [*********] on behalf of the Respondent at a Preliminary Meeting held on [*date*] at [*address*], I make the following orders:

1. As authorised by the parties I designate the seat of the arbitration to be London.

2. The parties' quantity surveyors are to meet forthwith as often as necessary with a view to narrowing the issues in respect of the Claimant's measured final account. If the parties are unable to agree whether or not a particular item in the final account submitted by the Claimant, or any part of any item, truly forms part of the measured work element of the Claimant's final account calculations or is, instead, a claim for loss and expense, it can be referred to me for decision as to whether or not it is measured work or loss and expense and, if it is measured work, for quantification under the procedure set out in paragraphs 3 and 4 below. These meetings are to be complete by [*date*] and by that date the quantity surveyors are to have agreed a list of major items upon which they disagree and which they wish to put to me for my decision.

3. I shall meet with the parties' quantity surveyors on a date no later than 14 days after the completion of the quantity surveyors' discussions, provisionally set for [*date[s]*], to consider

---

their respective views on each disputed item. No other representative of the parties shall be present at these meetings. These meetings are to include consideration of an appropriate method of dealing with those items where the amount in dispute between the parties is so small that individual consideration of each one by myself will not be a cost-effective procedure. Once I have reached a view as to the proper figure for the claimant's measured work final account I shall make a declaratory Award as to that figure.

4. Any right to payment in respect of the declared measured work final account figure will be the subject of separate submissions by the parties. A timetable for those submissions will be ordered by me at the appropriate time.

5. The Claimant is to serve its Statement of Claim on the Respondent on or before [*date*].

6. The Respondent is to serve its Statement of Defence and Counterclaim on the Claimant no later than [*date*] or twenty one days after the Claimant serves its Statement of Claim.

7. The parties' respective Statements are to set out the factual and legal basis relied upon and are to include copies of the principal documents which the party considers necessary to support its Statement.

8. The Claimant is to serve its Defence to Counterclaim on the Respondent no later than [*date*] or fourteen days after the Respondent serves its Statement of Counterclaim.

9. Any application to serve a Reply to the Defence or to the Defence to Counterclaim or to serve a Request for Further and Better Particulars of any Statement shall be made in writing and be fully reasoned within 14 days of service of the relevant Statement.

10. All Statements and Requests are to be copied to me.

11. The parties are to seek to agree a chronology of events no later than the service of the Defence to Counterclaim.

12. Statements should not be served on me by fax. If time constraints require service of a Statement on the other party by fax a covering letter only may be copied by fax to me if a party wishes to confirm that service has been achieved and the Statement sent to me by post thereafter.

13. Requests for specific discovery shall be made direct between the parties. Only where objection is taken to the nature of the request should there be any reference to me.

14. Documents submitted with Statements are to be deemed accepted by the other party and thus admissible evidence in this arbitration unless objected to at the time of or before service of the following Statement.

15. The parties are to agree a preliminary list of issues within 14 days of service of the last Statement.

16. Written statements of witnesses of fact shall be exchanged within 28 days of the service of the last Statement with copies to me. These statements are to stand as evidence in chief.

17. Experts of like disciplines shall exchange draft reports within 28 days of exchange of witness statements. Thereafter they shall meet on a without prejudice basis as often as necessary in order to agree facts and eliminate issues so far as possible. A joint report indicating those matters agreed and those disagreed giving reasons for the disagreement shall be completed within 28 days of the exchange of draft reports and shall stand as evidence in chief.

18. The site will only be viewed if considered necessary by myself or by either party after the hearing.

19. Legal submissions at any stage are to be reduced to writing.

20. A telephone conference call is to be arranged towards the end of [*month*] to consider the further conduct of this arbitration.

21. A period of four weeks for the hearing is provisionally reserved commencing on [*date*].

22. Communications to me are always to be copied to the other party and so endorsed.

23. The costs of the meeting and of this Order shall be costs in the Arbitration.

24. Liberty to apply.

Signed

[*Arbitrator's name*]
Arbitrator
[*date*]

To:

| [*Party representatives' name*] | Representatives of the Claimant |
| [*Party representatives' name*] | Representatives of the Respondent |

The initial order is a most important document that forms the framework for the conduct of the arbitration. It is worth the arbitrator's while considering obtaining the parties' agreement to a draft prior to finalising this order.

This example is applicable to a 'full' arbitration based on the pattern of litigation. There may be many more requirements discussed and fuller directions will then be needed. There may also be matters discussed which do not require orders but will need confirmation. These can be set out in the covering letter or in an addendum to the order.

Figure 2.10 is an example of a First Order where there has been no preliminary meeting.

FIGURE 2.10 – FIRST ORDER APPLICABLE TO A DOCUMENTS ONLY ARBITRATION WITH UNREPRESENTED PARTIES

In the Matter of the Arbitration Act 1996

and

In the Matter of an Arbitration

between

[*name of Claimant*]   CLAIMANT

and

[*Name of Respondent*]   RESPONDENT

ORDER No. 1

Following correspondence between the parties and myself and with the agreement of the parties I make the following orders.

1. As authorised by the parties I designate the seat of the arbitration to be London.

2. Each party shall prepare a full Statement of Case setting out his view of the circumstances leading up to and the details of the dispute between the parties including a statement of any monetary or other recompense sought. Each Statement shall be accompanied by copies of all documents upon which the party serving the Statement seeks to rely in support of the statements and submissions made. If a party wishes to include the evidence of witnesses in support of his case, this must be in the form of a written statement of his evidence by the witness himself.

    The Claimant shall serve his Statement of Claim no later than [*date*].

The Respondent shall serve his Statement of Defence and Counterclaim no later than [*date*] or [*number*] days after the Claimant serves his Statement of Claim.

The Claimant shall serve his Statement of Defence to Counterclaim no later than [*date*] or [*number*] days after the Respondent serves his Statement of Counterclaim.

A copy of each Statement and all supporting documents shall be sent to me at the time of service.

3. I attach a copy of an extract of Guidance Notes prepared by the Chartered Institute of Arbitrators for an arbitration scheme which they administer which will, I suggest, assist you in preparing your statements and avoid some of the common pitfalls that lie in wait for those unfamiliar with preparing submissions for an arbitrator.

4. At my discretion I may call a meeting of maximum one day in length in order for any clarifications or additional statements to be made. Each party shall have the opportunity at the meeting to question the other party and any witnesses from whom the other party may have obtained statements. Each party shall have a maximum of two hours for the purpose of amplifying its case and posing questions of the other and allowing for procedural matters and any questions that I myself may have, the meeting should take no more than five hours. Any witness who has provided a statement should come to the meeting. An unsupported witness statement will not be ignored but the weight that I shall give to statements that there has been no opportunity to test by oral questioning will be limited.

5. In the interest of cost the parties shall present their cases personally.

6. The meeting shall take place at the premises that are the subject of the dispute on [*date*] to enable me to inspect the works as necessary.

7. Letters to me are always to be copied to the other party and this should be noted on each letter.

8. The costs of this Order shall be costs in the arbitration.

9. Either party may apply to me for additional or varied directions.

[*name*]
Arbitrator
[*date*]

To [*name*]   CLAIMANT

     [*name*]   RESPONDENT

It is worth emphasising that unrepresented parties often have difficulty in presenting a coherent case. An arbitrator could not be criticised for trying to help in this situation and the circulation of published guidance may well be appropriate.

Figure 2.10A is a typical set of published guidelines.

FIGURE 2.10A – EXTRACTS FROM GUIDANCE NOTES OF CHARTERED INSTITUTE OF ARBITRATORS

GUIDANCE NOTES FOR CLAIMANTS

1. Your statement of claim should set out in chronological order the events which have led to the claim and refer to each supporting document in respect of each allegation. The information included in the statement should include references to:

    (i)   the relevant parts of any contract;
    (ii) any specific requirements;
    (iii) what was promised and what was received;
    (iv) relevant dates;
    (v)   the names of persons concerned (e.g. Respondents, employees or agents),
    (vi) the amount(s) claimed, clearly and precisely quantified;
    (vii) the remedies sought, whether compensation, specific performance (i.e. performance or completion of a contract) or corrective works.

2. Claimants must prove their case. Where appropriate each allegation or element of the claim should be set out in the form of a table. This will help the Arbitrator to appreciate what the differences between the Parties really are.

3. You should avoid merely sending a bundle of all documents in your possession and calling this bundle 'the claim'. The 'claim' is the written statement of claim and the bundle of documents is the supporting evidence.

4. You should avoid 'dressing up' or exaggerating claims to make weight. Allegations which are not supported by evidence will not assist your case and may in fact damage it.

5. It is not enough for you merely to infer that your dealings with the Respondent led to loss, you must show that there was a breach of some term of the contract, express or implied, and that where appropriate, you took steps to reduce your loss. Allegations must therefore be set out precisely.

6. When making comments on the Respondent's defence, you should restrict them to the matters dealt with in the defence and

---

not raise any new points. You must also deal with any counter-allegations made in the defence.

7. If you are represented by a lawyer or other professional adviser, you should communicate with the CIArb or the Arbitrator only through your adviser. If you have appointed an adviser, direct communication may cause unnecessary work and thus delay.

GUIDANCE NOTES FOR RESPONDENTS (Also applicable to a Claimant preparing a defence to a counterclaim)

1. The defence should answer each and every point raised by the Claimant clearly and precisely, giving details of any action taken to remedy defects or reduce losses. Remember, any points in the claim which remain unanswered, will normally be treated as having been admitted.

2. Any counterclaim must be supported by all relevant evidence and must be clearly and precisely quantified.

3. You do not have an automatic right to reply to the Claimant's reply to your defence. If the Claimant raises new points or makes allegedly scandalous comments, you may be permitted to submit a reply to those comments. In the first instance, if you wish to reply, you must apply to the Arbitrator in writing, who will decide whether you may do so.

CHECKLIST FOR BOTH CLAIMANT AND RESPONDENT

You should ask yourself the following questions before submitting your statement of claim, defence or counterclaim.

1. Exactly what am I claiming for or defending against?
2. Why do I believe that something is due to me or nothing is due to the other party?
3. Have I clearly answered all points in the other party's arguments with which I disagree?
4. Am I submitting all the documents that the Arbitrator will need to consider? If in doubt about something, include it.
5. Have I complied with the time limits?

The following points should be noted:

-   It is essential to record that the directions are made with the consent of the parties or their representatives if this is the case.
-   It is usual to record the date and place of the preliminary meeting and who was present or the fact that the directions are based upon some agreement made by correspondence.

- The 'statements' are the formal documents in which the parties set out their cases and the timetable should be a realistic one. See subsequent discussion.
- The sample order (Figure 2.9) set out above includes a typical order relating to the process of discovery. In arbitration disclosure of documents is governed by section 34(2)(d) and the arbitrator may decide that it is far more appropriate to control discovery himself and only agree to the disclosure of individual documents which are specifically requested by one or other party. The question of discovery is discussed in more detail later.
- It will save time and cost if the parties can agree as many facts and figures as possible. This does not mean that they agree that an amount is due, but merely what that amount is if it is found to be due, e.g. the respondent may argue that the claimant is not entitled to any sum as compensation for delay at all, but that, if the arbitrator finds that he is entitled to it, then he agrees the correct figure would be £x.
- It is essential that there should be no communication with the arbitrator by either party without the knowledge of the other. Telephone calls should be avoided save on purely administrative matters such as dates and times of meetings where the arbitrator may arrange them by telephone with both parties and set out the final agreement in writing. Conference telephone calls are however most useful and can be used for wider purposes.
- If the parties propose to call expert witnesses it is customary to limit the number on each side and require exchange of their reports well before the hearing. Fig 2.9 includes a process whereby the experts produce a joint report which stands as the evidence in chief of both of them, thus dispensing with individual reports. Unless the parties have agreed otherwise the arbitrator has, under section 37(1)(a)(i), the power to appoint his own expert(s). The arbitrator can do this without reference to the parties although it would be sensible of him to inform them before he does so. It would be an unnecessary additional cost burden on the arbitration and possibly an additional problem for the arbitrator to resolve if the parties also appointed their own experts to deal with the same matters and he ended up with three differing expert views on the same subject, possibly even holding a fourth himself!
- 'Liberty to apply' means that either party can go back to the arbitrator and ask him for an amending order, e.g. extension of time for delivery of statements.

- It has been customary, if the parties are represented at the preliminary meeting by barristers – which is comparatively unusual – for the arbitrator to mark his directions 'Fit for Counsel' so that the barristers' fees can be taken into account when costs are assessed. As the 1996 Act requires the arbitrator to assess the costs himself, rather than a taxing masker, this is not really necessary any longer so long as he keeps adequate notes
- The order must record how the costs of the meeting and the order are to be dealt with. 'Costs in the arbitration' means that the costs involved will be dealt with in the eventual award.

The examples should not be followed slavishly since the order for directions must always embody what was agreed – or decided by the arbitration – and various problems can arise:

- One party may try to insist on full formal pleadings and what in effect is a procedure analogous to that of the Court. This procedure is only strictly necessary or desirable where there are several heads of claim, each with a different legal basis, and each one of which is linked to the eventual amount of the award.
- It may not be clear which party is to be treated as claimant and which as respondent. This situation may arise where there are both claims and counterclaims and where oral agreements are wholly or partly relied on. The starting point must be 'Who sought arbitration first?' The answer to this question may not be decisive where, for example, one party has commenced ligitation and the court has ordered a stay, and in such a case it is probable that the person who controls the major claim is properly to be regarded as the claimant.

These are merely examples of the sort of problems which can arise at this stage and hence of the need for the arbitrator to have a detailed knowledge of procedure and ready access to the standard works of reference. The arbitrator should never be rushed into making a decision, however hard he may be pressed to do so. If in doubt, the best advice is for the arbitrator to adjourn the meeting and take time to research and consider the point. He must always give both parties the opportunity to make full submissions to him wherever he is asked to make a ruling on any matter. The arbitrator should resist the temptation to respond directly to a submission before he hears the other party's point of view even if he feels that the original submission has absolutely no merit whatsoever. The arbitrator who falls into this trap may find himself in receipt of a response that needs an answer especially where the party making the submission is unrepresented. He is then in danger of getting himself ever deeper into the detail of the argument and then losing the somewhat detached approach that an arbitrator must retain. He should hold onto the concept that he is there to resolve the arguments between the parties even if this may mean waiting or even prompting a response from the other party. Very often a problem that appears insoluble at first glance will disappear once a response is received from the other party.

In the past it has been customary to suggest that it is inappropriate for an arbitrator to proceed with a preliminary meeting where one party fails to attend. The 1996 Act now empowers an arbitrator 'to proceed with an oral hearing of which due notice was given' when a party fails to attend without showing sufficient cause [section 41(4)]. The arbitrator will have given notice of the preliminary meeting and this meeting is by its very nature procedural in content. The order that the arbitrator produces will be endorsed with 'liberty to apply' and this will enable the absent party to put forward any points that it may wish to and the arbitrator, after due consideration and allowing the other party to comment, will, if he thinks fit, be able to amend his directions. He should therefore seriously consider proceeding in any event even though he should make some effort to ascertain whether the absent party has good reason for not being there before doing so.

2.5 Identifying the issues

The arbitrator has to know what it is that the parties want him to resolve. The issues have to be identified. This can be done by either 'Statements of Case' or traditional 'Pleadings' as used in the courts.

In arbitration it is common to order statements of case although pleadings are not unknown. The purpose of this documentation is to set out clearly, and as concisely as possible, the case that each party wants to put forward. Each case must be set out fully so that the other party may know what it is and have every opportunity of dealing with it. The arbitrator cannot make a decision on any matter which is not set out in the pleadings or statements of case and this is why these documents are often framed in the alternative, e.g. 'I am not liable at all, but, if I am liable, I am not liable for as much as the claimant says I am'.

It is generally accepted in arbitration that the preparation of statements of case is preferable to the more traditional pleadings. A pleading is a statement of those matters which the party preparing the pleading intends to bring evidence to prove. There is no requirement for any indication of the legal aspects of the case. A statement of case, whether it refers to the claim or the defence, is normally prepared in narrative style and sets out both matters of fact and of law and will often include evidence such as documents and statements from witnesses. By its nature a full statement of case will assist the other party to understand the contentions fully and it is seen as a method of encouraging settlement. The reverse of the coin is that a statement of case takes considerably more time to prepare properly than it takes to prepare a pleading. This means that the costs of the parties can be considerably inflated at the outset at a stage where the resolution of the dispute may be better served by a pleading which identifies the issues but goes little further. There are arguments for and against each approach and it is for the arbitrator, in discussion with the parties at the preliminary meeting to decide the most appropriate and it is to be hoped ultimately the most cost effective approach.

Formal pleadings consist of:

* Points of claim
* Points of defence and of counterclaim (if there is one)
* Points of defence to the counterclaim (if there is one)

It has been customary in the past to allow the parties to make a formal 'Reply' to a defence, and in some circumstances a reply to a defence to a counterclaim. It is however notable that section 34(2)(c) refers only to written statements of claim and defence. It is therefore suggested that the parties should be required to make a reasoned application for leave to serve any such reply. The reasoning is that each party should put its case definitively once only and there should be no need, if statements of claim and defence are properly drafted, for a second bite at the cherry. It should be borne in mind in this respect that unrepresented parties will seldom get it right first time and some flexibility is often necessary even to the extent in some cases of allowing the respondent to make a reply to the reply.

Statements of case proceed in a similar way but are generally entitled 'Statements' rather than 'Points'.

Figures 2.11 and 2.12 are examples of a statement of claim and a statement of defence and counterclaim.

FIGURE 2.11 – STATEMENT OF CLAIM

In the Matter of the Arbitration Act 1996

and

In the Matter of an Arbitration

between

[Name of Claimant]	Claimant

and

[Name of Respondent]	Respondent

STATEMENT OF CLAIM

1. This Arbitration concerns a dispute between the Claimant, [*name and address*] as Contractor and the Respondent, [*name and address*] as Employer under a contract for the erection of a commercial office building and ancillary works at [*location*].

2. A copy of the contract is appended as Appendix 1. In addition to its express terms the Claimant maintains that the following terms are necessarily to be implied as terms of the contract in order to give it full business efficacy:
    1. That the Respondent and/or the Architect would not hinder or prevent the Claimant from carrying out and completing the contract Works in accordance with the terms and conditions of the contract; and

    2. That the Respondent and/or the Architect would do all that was reasonably necessary to enable the Claimant to carry out and complete the contract Works in accordance with the terms and conditions of the contract.

3. The claim comprises:
    1. Claims for entitlement to extensions of time under clause 25 of the contract over and above those already granted by the Architect, the last such extension expiring on or after the date of Practical Completion as certified by the Architect;
    2. A consequential claim for repayment of liquidated and ascertained damages which the Claimant maintains have been wrongfully deducted by the Respondent from payments due to the Claimant;
    3. A claim that, due to certain of the events set out below, the Completion Date under the contract is unenforceable against the Claimant who is therefore entitled to a reasonable time in which to complete the Works, and the

Claimant maintains that that reasonable time expired on or after the date of Practical Completion as certified by the Architect;

4. A consequential claim for a declaration that the Respondent is not entitled to any deduction or payment of liquidated and ascertained damages under clause 24.2 of the contract in any event and for reimbursement of all such damages deducted;

5. Claims for direct loss and/or expense under clause 26 of the contract;

6. In relation to certain events set out below a claim in the alternative for damages for breach of the express and/or implied terms of the contract by the Respondent and/or the Architect on its behalf.

4. Appendix 2 to this statement contains a list of all the principal documents relied upon in support of this statement and copies of all documents expressly referred to in this statement in chronological order. Each page of the copies of documents has been separately numbered for ease of reference. Where an individual passage only has been referred to a vertical line has been drawn against that passage. Numbers in bold type and in square brackets thus **[21]** in the text below are references to the relevant page numbers in the copies of documents in Appendix 2.

5. As will be seen the Date for Completion in the Appendix to the contract was 30 September 1996. During the course of the Works the Architect issued five extensions of time in relation to certain delaying events **[46, 51, 75, 87, 105]**. These extensions of time are not in themselves contested and the events for which they are granted do not appear in the narrative which follows except as is necessary in relation to other events.

6. The Completion Date fixed by those extensions of time was 31 January 1997. Practical Completion of the Works was certified by the Architect to have been achieved on 27 June 1997, and this is not contested by the Claimant.

7. There follows a complete narrative of events on this contract which have given rise to this claim. Full details of the claims made in respect of each event are set out in the Schedules annexed to this statement. These include, where necessary, comparative programme and progress charts showing the effect of each event upon the progress of the Works and, where relevant, upon the completion of the Works. References to clause numbers are to clauses of the Conditions of Contract.

8. **Event No 1: Delay in provision of information relating to foundations**

8.1 On 9 January 1995, three weeks before the Date for Possession of the site of the Works, the Claimant sent to the Architect the master programme for the Works as required by clause 5.3.1.2 **[5–7]**, and a detailed schedule of information required relating to the foundations with the dates when each item of information was required if the programme for the foundations works was to be achieved **[8]**.

8.2 The schedule required full details of reinforcement for the pile caps and connecting beams, including bending schedules, to be provided by 30 January 1995. This was absolutely necessary since the reinforcement fabricators required three weeks in which to carry out the fabrication and deliver the reinforcement to site for fixing. Drawings showing the reinforcement were supplied by 30 January 1995, but the bending schedules were sent in instalments extending from 5 February 1995 to 20 February 1995. As a consequence, the reinforcement had to be ordered in batches as the schedules were received and passed immediately to the fabricator, and the last reinforcement was not delivered to site until 13 March 1995, one week after the programmed date for completion of the pile caps and beams. Allowing for the fixing of the reinforcement and the inevitable disruption caused to the casting of the pile caps and beams consequential upon the uncertainty as to when and in what sequence the bending schedules would be received, the pile caps and beams were not completed until 31 March 1995, 25 calendar days after the programmed date.

8.3 The claimant wrote to the Architect on 30 January 1995 **[9]** as soon as it became apparent that the bending schedules would not be received by that date, pointing out that this would inevitably cause delay and disruption to the Works and giving the notice required under clause 25.2.1.1 of a Relevant Event under clause 25.4.6 and making an application for reimbursement of the direct loss/and or expense which would inevitably be incurred under clause 26.1. The Architect responded on 3 February 1995 **[10]** denying that the bending schedules were necessary for the ordering of reinforcement and that any delay or disruption would inevitably result from any delay in their provision. Further correspondence followed **[11–17]**. The Claimant wrote to the Architect on 13 March 1997 **[18]** setting out its estimate of the delay in completion of the Works as required by clause 25.2.2, and on 24 April 1995 **[19–29]** giving details of the loss and/or expense incurred.

8.4   No extension of time has been granted and no direct loss and/or expense has been certified or received in respect of this event. Details of the delay and disruption and the direct loss and/or expense incurred are set out in Schedule 1. The Claimant claims entitlement to an extension of time of 25 calendar days and direct loss and/or expense as now fully particularised in Schedule 1 in the sum of **£17,895.46** plus interest thereon from 15 May 1995, the day by which payment should have been received under the first certificate for interim payment issued after 24 April 1995, to the date of the Award.

9.   **Event No 2: Variations in ground floor slab service ducts**

9.1   ..... (etc.)

SUMMARY

In summary the Claimant claims:

1.   Extensions of time up to 27 June 1997 or later;

2.   Direct loss and/or expense in the total sum of £297,672.87.

3.   Alternatively direct loss and/or expense in the total sum of £250,486.72 and damages in the total sum of £47,186.15.

4.   Reimbursement of liquidated and ascertained damages deducted by the Respondent from payments due in the total sum of £210,000.00.

5.   Interest under section 49(3) of the Arbitration Act 1996 at the rate of 4% above Midland Base lending rate compounded monthly for the periods set out in respect of each event.

6.   Recoverable costs under section 63 of the Arbitration Act 1996.

7.   Interest under section 49(4) of the Arbitration Act 1996 at the rate of 4% above Midland Bank base lending rate compounded monthly on all sums directed to be paid in any Award including interest and costs from the date of the Award until payment.

SERVED BY [*name*] ON BEHALF OF THE CLAIMANT ON [*date*]

FIGURE 2.12 – STATEMENT OF DEFENCE AND COUNTERCLAIM

In the Matter of the Arbitration Act 1996

and

In the Matter of an Arbitration

between

[*Name of Claimant*]   Claimant

and

[*Name of Respondent*]   Respondent

STATEMENT OF DEFENCE AND COUNTERCLAIM

Note: unless otherwise stated all references to paragraph numbers are to the paragraphs in the statement of claim.

1. Paragraphs 1 and 2 are agreed.

2. Appendix 1 to this statement corresponds to Appendix 2 to the statement of claim and is similarly arranged and referenced. For ease of reference all documents specifically referred to in this statement are listed and copies included including those listed and included in Appendix 2 to the statement of claim.

3. Paragraphs 5 and 6 are agreed.

STATEMENTS OF DEFENCE TO THE CLAIM

4. There follows the Respondent's response to the narrative statement in relation to each event set out in the statement of claim. Without prejudice to denials of liability in respect of each event, a response to the particulars set out in each Schedule to the statement of claim is provided in the corresponding Schedule to this statement.

5. **Event No 1: Delay in provision of information relating to foundations**

5.1 Paragraph 8.1 is admitted.

5.2 It is admitted that the schedule of information required full details of reinforcement for the pile caps and connecting beams to be provided by 30 January 1995. However the schedule made no specific reference to bending schedules, and it is denied in any case that their provision by the date stated in the schedule was necessary for reinforcement to be supplied by the dates required in order to meet the programme for completion of the

66

pile caps and beams. The reinforcement could have been ordered from the details on the Consulting Engineer's drawings, all of which had been provided to the Claimant by 30 January 1995. It is agreed that the pile caps and beams were not completed until 31 March 1995, but it is denied that this was due to any delay in the provision of bending schedules. On the contrary it is believed by the Respondent, without any admission of any need for the Respondent to prove the true cause, that the delay was due to problems which the Claimant had with the supplier of reinforcement, some of which it is understood had to be returned to the supplier for rebending due to errors made by the supplier.

5.3   Paragraph 8.3 is admitted.

5.4   As to Paragraph 8.4, it is denied that the Claimant is entitled to any extension of time or to any direct loss and/or expense in relation to this event. However without prejudice to this assertion, the Respondent contests the particulars given in Schedule 1 to the statement of claim. If, which is denied, the Claimant is entitled to any extension of time and/or any direct loss and/or expense in relation to this event the period and amount to which it is entitled are as set out in Schedule 1 to this statement.

6.   **Event No 2: Variations in ground floor slab service ducts**

6.1   ... (etc.)

STATEMENT OF COUNTERCLAIM

1.   The Respondent maintains its entitlement to retain the sum of £210 000 as liquidated and ascertained damages under clause 24.1 of the contract.

2.   Further the Respondent counterclaims for defects in the Works which remain unremedied by the Claimant and the date of this statement as follows.

2.1   Tests have shown that the concrete used in the construction of the first floor of the building is not of the quality or strength specified in the contract. Test certificates are included in Appendix 1 hereto **[210–215]**. As a consequence it has not been possible fully to let the first floor of the building for the purposes for which it was intended. Details of the consequential loss incurred by the Respondent together with the cost of the tests are set out in Schedule 17 hereto. The Respondent claims the sum of £215 671.00 as damages in respect of the Claimant's breach of contract in this respect.

2.2   ... (etc.)

SUMMARY

In summary:

1.   The Respondent denies all claims made in the statement of claim.

2.   The Respondent maintains its entitlement to retain the sum of £210 000 as liquidated and ascertained damages under clause 24.1 of the contract.

3.   The Respondent counterclaims the sum of £576 726.89 in respect of the matters set out in the statement of counterclaim.

4.   The Respondent claims interest under section 49(3) of the Arbitration Act 1996 at the rate of 4% above Royal Bank of Scotland base lending rate compounded monthly as appropriate in respect of all sums not received or expended.

5.   The Respondent claims its recoverable costs under section 63 of the Arbitration Act 1996.

6.   The Respondent claims interest under section 49(4) of the Arbitration Act 1996 at the rate of 4% above Royal Bank of Scotland base lending rate compounded monthly on all sums directed to be paid in any Award including interest and costs from the date of the Award until payment.

SERVED BY [*name*] ON BEHALF OF THE RESPONDENT ON [*date*]

If statements of claim or defence are properly prepared, there should be no need for any further explanation or elaboration. It may, however, happen that one party genuinely needs further explanation of the other party's case in order to respond to it adequately. In that case the party concerned should be required to make a reasoned application for leave to make what is called a 'Request for Further and Better Particulars'. If leave is given time limits should be set for the service of the request (unless, as ideally should be the case, the request is attached to the application for leave) and for service of the particulars requested. The arbitrator should however be aware that requests of this nature can be used as a delaying tactic and should be on his guard for such a situation.

Figure 2.13 is a specimen request for further and better particulars; Figure 2.14 is a specimen reply.

FIGURE 2.13 – REQUEST FOR FURTHER AND BETTER PARTICULARS

In the Matter of the Arbitration Act 1996

and

In the Matter of an Arbitration

between

[*Name of Claimant*]   Claimant

and

[*Name of Respondent*]   Respondent

**REQUEST FOR FURTHER AND BETTER PARTICULARS
OF THE STATEMENT OF CLAIM**

1.   **Of paragraph 3** please state all facts and matters relied upon in support of the assertion that the terms stated are to be implied in the Agreement to give full business efficacy thereto.

2.   **Of paragraph 5** please state the manner by which the alleged deferment of possession of the site was conveyed to the Claimant. If by statement at a meeting please state the date, time and location of the meeting, who was present and by whom and in what manner the statement was made; if by letter please state the date and reference of the letter.

3.   **Of paragraph 8** please state

> 1. in respect of each notice of delay the manner in which the notice was given, the form of the notice and to whom it was addressed.
> 2. in respect of each application the like particulars as in respect of notices under sub-paragraph 1.

SERVED BY [NAME] ON BEHALF OF THE RESPONDENT ON [DATE].

Quite commonly, parties ask for an extension of time for service of the various documents and such requests are dealt with on their merits by the arbitrator. It is not usual to refuse such a request in an arbitration but if the arbitrator considers that there is not sufficient reason, or the request is a delaying tactic, he should warn the party concerned that the delay will be taken into account in any award of costs.

FIGURE 2.14 – REPLY TO REQUEST FOR FURTHER AND BETTER PARTICULARS

In the Matter of the Arbitration Act 1996

and

In the Matter of an Arbitration

between

[*Name of Claimant*]   Claimant

and

[*Name of Respondent*]   Respondent

REPLY TO REQUEST FOR FURTHER AND BETTER PARTICULARS OF THE POINTS OF CLAIM

1.   Request

Of paragraph 3 please state all facts and matters relied upon in support of the assertion that the terms stated are to be implied in the Agreement to give full business efficacy thereto.

Reply
The Claimant relies upon the statement in the Statement of Claim. No further particulars will be given.

2.   Request
Of paragraph 5 please state the manner by which the alleged deferment of possession of the site was conveyed to the Claimant. If by statement at a meeting please state the date, time and location of the meeting, who was present and by whom and in what manner the statement was made; if by letter please state the date and reference of the letter.

Reply
By statement at the meeting referred to in paragraph 5 of the Statement of Defence. It was made by the Employer in the words: 'I am afraid I cannot give you possession of the site on 18 March after all. Would 4 April suit you?' or words to the like effect.

3.   Request
Of paragraph 8 please state
1.   In respect of each notice of delay the manner in which the notice was given, the form of the notice and to whom it was addressed.

> 2. In respect of each application the like particulars as in respect of notices under sub-paragraph 1.
>
> **Reply**
> 1. By letters from the Claimant to the Architect dated [*list of dates*]
> 2. By letters from the Claimant to the Architect dated [*list of dates*]
>
> **SERVED BY [NAME] ON BEHALF OF THE CLAIMANT OR [DATE]**

2.6 *Discovery and inspection*

Discovery of documents 'is the procedure under which one party discloses to the other the documents which he will produce at the hearing' (Powell-Smith and Chappell, *Building Contract Dictionary*, 1985, p 169). Inspection is the process of looking at the documents disclosed. In English law each party must disclose to the other all the documents in his possession and/or control which are relevant to the issues, even though these may be unfavourable to his case. Indeed, both parties may be required to swear affidavits stating that they have made full disclosure of all documents, though this is only done in arbitrations if there is some dispute over the matter.

In litigation the parties will exchange lists of the documents in question. All documents must be included on the list, even those for which what is known as 'privilege' is claimed, which means that the document is one which by law the party is not bound actually to show to the other party, e.g. opinions of counsel about the case and correspondence with his own solicitor.

Section 34(2)(d) of the 1996 Act deals with the disclosure of documents in arbitrations. By this section the arbitrator, unless the parties agree otherwise, has complete control over which documents or classes of documents should be disclosed. Full discovery, as it is known in litigation, does not apply as of right in arbitration. One of the results of this is that the traditional discovery process and the cost and time resulting therefrom can be avoided. Of course if the parties want it they can have full discovery. Arbitration can in this respect be a real Rolls-Royce of dispute resolution procedures, with the associated cost! It is more usual however that the parties seek to keep costs down by restricting discovery. The arbitrator will generally not make a unilateral decision in this respect, he will invite the parties to address him as to the appropriate extent of discovery.

The use of a statement of case procedure will very often mean that all the documents that a party wishes to rely on are included with the statement. As mentioned above the cost of this can be considerable. Extra time may also be needed to research all the documents relating to a contract, which can be considerable in number, to find those that relate to the dispute. The option that is sometimes used with a statement of case procedure is to produce the narrative statement together with the legal argument and limit the documents to those that are considered to be the most important. This does mean that there may need to be a more exhaustive discovery process later on and the arbitrator will ultimately have to decide on the extent of this.

The format of a traditional list of documents is shown as Figure 2.15.

FIGURE 2.15 – LIST OF DOCUMENTS

In the Matter of the Arbitration Act 1996

and

In the Matter of an Arbitration

between

[*Name of Claimant*]   Claimant

and

[*Name of Respondent*]   Respondent

CLAIMANT'S LIST OF DOCUMENTS

The following is a list of documents relating to the matters in dispute in this arbitration which are or have been in the possession or control of the Claimant and is served pursuant to the Arbitrator's Order dated [*insert date*].

1. The Claimant has in its possession or control those documents enumerated in Part 1 of Schedule 1 hereto which are original documents.
2. The Claimant objects to the production of the documents enumerated in Part 2 of Schedule 1 hereto on the ground that they are by their nature privileged.
3. The Claimant has had but has not now in its possession or control the documents enumerated in Schedule 2 hereto.
4. Neither the Claimant nor its advisers nor any other person on its behalf have now or have ever had in their possession or control any documents of any description whatsoever relating to any matter in dispute in this arbitration other than the documents enumerated in Schedules 1 and 2 hereto.

SCHEDULE 1 PART 1

Bundle A
Articles of Agreement and Conditions of Contract
Bills of Quantities Volumes 1 and 2
Contract Drawings

Bundle B
1. Letter from the Architect to the Claimant dated [*insert date*]
2. Copy of letter from the Claimant to the Architect dated [*insert date*]
3. Copy of letter from the Architect to the Respondent dated [*insert date*] with compliments slip
[etc.]

Bundle C
Interim Valuations comprising pages 1 to 357 inclusive

Bundle D
Minutes of site meetings comprising pages 1 to 260 inclusive
[etc.]

SCHEDULE 1: PART 2

Correspondence between the Claimant and its Solicitor.
Advice of Counsel relating to matters in dispute in this Arbitration.
'Without prejudice' correspondence between the Claimant and the Respondent.

SCHEDULE 2

The originals of those documents listed in Part 1 of Schedule 1 which are copies of documents originating from the Claimant.

NOTICE TO INSPECT

Take notice that the documents listed in Part 1 of Schedule 1 may be inspected at the office of the Claimant's Solicitors during normal office hours by prior appointment following the notice specified by the Arbitrator in his Order for Directions.

BY [NAME] ON BEHALF OF THE CLAIMANT ON [DATE].

Once the lists of documents have been exchanged the parties, or more usually their solicitors, will meet and inspect each other's documents, taking copies as required. It is usual and desirable for them to prepare an agreed bundle of those documents which each side will rely on in evidence. The advent of photocopying means that the parties tend to produce copies of all the documents in their files because it is cheaper to do that than to select the ones they really need. This practice should be discouraged, as it leads to very large bundles of documents and in fact it probably wastes more in costs at the hearing than it saves at any other time. The arbitrator should insist that the bundle is in chronological order, consecutively numbered and preferably indexed, and that all photocopies are clear and legible.

Some arbitrators prefer to have the bundle of documents delivered before the hearing so that they may study them in advance. Others prefer to have the bundle available at the beginning of the hearing so that they may be specifically referred to the documents which the parties wish to draw their attention to. It is a matter of personal preference, but the arbitrator should always, at some stage before making his award, read all the documents presented to him, because they are by definition all part of the evidence. It sometimes happens that when reading the documents after the hearing, the arbitrator will come across some point affecting the claim or counterclaim which has not been put to him at the hearing. If the parties were unrepresented at the hearing, it is suggested that he should then reconvene the hearing so that he may put the point to the parties and hear any argument, since it would be improper either for him to decide the case on a point not argued, or if he otherwise fails to give both parties an opportunity of arguing the case. Sometimes parties do not agree on the documents and each will then present a separate bundle. In that case they must each 'prove' the documents. For instance, the writer of a letter may produce a copy of it and the other party may deny receiving the original or, in an extreme case, state that the copy is a forgery made up after the event. In that case the writer will be required to state on oath that it is a true copy of a letter written by him which was genuinely dispatched.

2.7 Matters of law

Most technical arbitrators are in general more comfortable dealing with matters of law on the basis of written submissions. It is the exceptional technical arbitrator who has a sufficient grasp of the law to do full justice to oral submissions from counsel. During such oral submissions he will have to take exhaustive notes. He may find that on examination of these notes he still does not understand the submissions properly and he may have to reconvene to hear further argument or he may decide to require written submissions both of which waste considerable time and increase the cost of the arbitration. In general the technical arbitrator would do well to require that all legal submissions be made initially in writing.

2.8   *Witnesses of fact and experts*

Current practice is for each witness to provide a written statement which sets out his evidence. The majority of arbitration hearings are held many months if not years after the event. It is often the case that a witness in preparing this statement will unintentionally, or even perhaps intentionally, embroider his evidence in the hope of assisting the party on whose behalf he is giving evidence. The arbitrator must as a result provide the other party with the opportunity of probing each witness's evidence for inaccuracies. This is done by cross examination. The time spent at the hearing is much reduced when the evidence in chief is given in writing. There is however the problem that the side producing the witness does not necessarily know what particular issues the other side is going to home in on. As a result a witness statement can be exceedingly lengthy as it has to cover everything 'just in case' and costs can be much inflated as a result. There is no real happy medium. Some arbitrators require the parties to limit the witness statements to cover major points only and allow limited further evidence in chief during the hearing if a party considers it to be appropriate.

In the case of experts the only really effective way of dealing with their evidence is by the production of reports. An expert is, after all, giving evidence of opinion and he must be given the chance to develop his contentions in writing. This procedure also has the benefit that the experts are likely to identify matters that they are in agreement upon. By ordering meetings of experts and the production of an agreed report setting out those matters that the experts agree and those that they do not the arbitrator can make a real contribution to the reduction in the cost of the hearing and thus the overall cost of the arbitration. As noted above, the possibility exists to dispense with final reports entirely. It will be possible then to require the opposing experts to be examined simultaneously at the hearing.

Figure 2.16 is a checklist of the headings in an expert's report.

FIGURE 2.16 – CHECKLIST OF HEADINGS IN AN EXPERT'S REPORT

- Name, address, qualifications and experience of the expert
- Description of instructions, terms of reference, issues required to address
- Documents and other exhibits examined
- Inspection of the subject-matter of the dispute, description of the technical investigation or enquiry
- Facts noted from the documents, exhibits and inspection
- Conclusions drawn from those facts with reasons

2.9 Preparation before the hearing

As soon as it is possible to fix the dates for the hearing, the arbitrator should do so in consultation with the parties and their advisers. In a large-scale arbitration much will depend upon the availability of counsel. Preferably the venue should be impartial (e.g. the arbitrator's own office, a hotel or other place where rooms are available for hire), but there is often no objection to using a room at the offices of one of the solicitors or, if the local authority is involved, at the council offices. In small arbitrations (e.g. NHBC arbitrations), if is often convenient to hold the hearing at the site so that an inspection can be made of any disputed defects, etc. at the time.

In general, if it is necessary to book a room for which hire charges are payable, the arbitrator would be well advised not to assume responsiblity for payment himself. The usual practice is for the venue arrangements to be made by the claimant. If there is difficulty in finding a venue the arbitrator may be able to help, for instance by finding rooms at his professional institution.There are a number of 'Arbitration Centres' in London and elsewhere where rooms can be hired.

In larger arbitrations the arbitrator will call a 'pre-hearing review meeting' probably four to six weeks before the hearing starts. The initial purpose of this meeting is to check that there is nothing left to do in respect of the exchange of statements of case or any further and better particulars that may have been requested, that discovery has been completed and that the statements and reports of the witnesses have been exchanged. He will also check that any directions that he may have given during the period since the preliminary meeting have been complied with. More importantly he must check that a suitable venue has been arranged and that all the other preparations for the hearing are in place.

For instance, he may, if he hasn't done so already, require the parties to provide him and each other with lists of legal authorities to which they intend to refer and to provide, at the hearing, photocopies of those authorities. This is especially important in relation to case law, and he should always require the parties to give a full copy of any case which is cited and not merely an extract which is favourable to them.

These matters on smaller arbitrations will generally be done by means of correspondence.

A point not to be forgotten is the layout of the room in which the hearing is to be held and the ancillary accommodation. Hotels are notoriously deficient in this respect. If a full formal hearing is required, there should be a separate retiring room for the arbitrator and, ideally, accommodation for the parties and their advisers so that they may hold private conferences. This ideal is seldom achieved unless a purpose-built room is used.

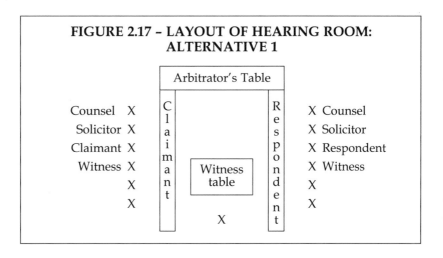

FIGURE 2.17 – LAYOUT OF HEARING ROOM: ALTERNATIVE 1

The hearing room itself can be set in various ways. A common method is to set the tables and chairs in a U shape (*Figure 2.17*); an alternative is to use the courtroom style layout, as illustrated in *Figure 2.18*. The parties, their witnesses and the arbitrator must have sufficient room both for themselves and for books and documents. In a building arbitration there are often many drawings to be consulted and these are best set out on a separate table. In many cases all these questions will have been decided at the preliminary meeting or at an earlier stage in the proceedings, and in general problems do not arise unless the parties are unrepresented.

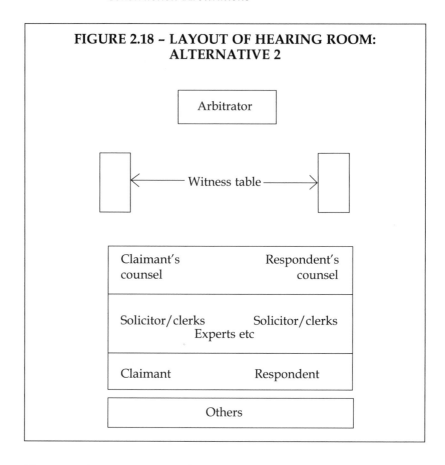

**FIGURE 2.18 – LAYOUT OF HEARING ROOM:
ALTERNATIVE 2**

There is of course no specific requirement as to which side of the room each party is placed. This is often done by the parties themselves before the arbitrator arrives.

Chapter Three
Problems Before the Hearing

3.1 Introduction

It is an old saying that there are no problems, only opportunities. This is quite true in arbitration. We have included the word 'problems' in the heading of this chapter but many of the situations described may well be better viewed as opportunities to overcome some of the difficulties that arise in any situation where there are two disputing parties.

Some typical situations that can arise at the time of the appointment itself were considered in chapter 2. In this chapter we look at a number of other situations which may arise between his appointment and the hearing and which may cause the arbitrator problems if he is not prepared for them.

3.2 Jurisdiction

It was suggested in Chapter 2 that the arbitrator must identify the extent of his jurisdiction at the preliminary meeting in order to ensure that he deals with all the matters that are referred to him and conversely does not deal with matters which are not within the compass of his appointment.

But what about the situation where a party, usually the respondent in the case of a unilateral application to an appointing authority by the claimant, suggests that the arbitrator has no jurisdiction whatsoever? A typical example is where, as is not uncommon, the contract documents are unsigned and the claimant has obtained the appointment of an arbitrator by sending the form of contract that he thinks applies to the appointing authority. The respondent immediately upon being notified of the appointment tells the arbitrator that the appointment has been made on the wrong form of contract and in fact, rather than the President of the RICS, it is the President of the RIBA who should have made the appointment.

Immediately the newly appointed arbitrator has a problem. He must first ascertain if the parties have reached an agreement under section 30 of the 1996 Act to the effect that he cannot rule on his own substantive jurisdiction. If they have reached such an agreement he is left with one problem only, does he proceed with the reference or not (section 32(4))? In the interests of costs it is likely that the arbitrator will not proceed and will await the decision of the court for this is the only place that the claimant can go in these circumstances for a decision on jurisdiction. If however the claimant is insistent that he proceeds, provided that he ensures that the claimant is fully aware of the risk that he is running, the arbitrator should proceed. He will of course want to ensure that his fees are secured in these circumstances.

A more likely scenario is where the parties have been at loggerheads for a long time. Why else was there an institutional appointment? There will then be no agreement under section 30 that will prevent the arbitrator from ruling on his own jurisdiction. He is in receipt of an appointment that has been made in good faith. How does the arbitrator sort this out?

The answer is that he must require submissions from both parties upon the objection. As the respondent is making the objection, he must make his submission first. The claimant must have the opportunity to answer. It is not obligatory that the respondent be given the opportunity to reply but it is often the case that the claimant may raise matters that are not included in the respondent's submission and the respondent must, in that situation be given the opportunity to reply. Sometimes even the respondent's reply results in the claimant suggesting that he must have the opportunity to make a further submission. If the arbitrator is not careful he will end up with submissions flying backwards and forwards with no seeming end in sight. In a situation where the claimant is adamant that he must reply, the best solution for the arbitrator is to require that the claimant satisfies him that there is good reason for a further bite of the cherry before permitting a reply.

The above scenario is not, of course, exclusive to jurisdictional problems and can occur at any stage of the proceedings.

The next conundrum for the arbitrator is whether to make a specific award on his jurisdiction (section 31(4)(a)) or to deal with the objection in an award on the merits (section 31(4)(b)). This is a decision that must be based upon the individual circumstances. In general, as discussed above it must be right that the arbitrator does not allow costs to be incurred unnecessarily. Preferably the jurisdiction question should be taken as a preliminary issue, dealt with by award and the rest of the arbitration put on hold until that award is made. Individual circumstances will however alter this and there is no hard and fast rule. If there were the alternatives would not be in the Act.

One further point that the arbitrator may need to consider is, what does he do if he finds that he has no jurisdiction? Should he write his award, operate the lien given to him by section 56 of the 1996 Act and only release it once paid? The answer is that, as usual, it depends on the circumstances. If the arbitrator has spent many hours over lengthy submissions there is no reason why he should not make an award and expect payment before releasing it.

But it may be a matter that is so obvious that the arbitrator sees almost upon opening the papers that he has no jurisdiction. Perhaps, as proposed above, the wrong appointing body has been used. In that instance rather than spending his time and the claimant's money writing out a formal reasoned award, the arbitrator would serve the resolution of the dispute best if he writes a short note informing the parties of his findings. It may be that the respondent will agree to accept the appointment, but it is rather more likely that, as objections are usually made by parties who are reluctant to see the arbitration going too fast, the claimant will have to start again with another appointing authority. In that event the arbitrator has, by his action, minimised one problem in the way of a party who is not going to have an easy ride in getting his dispute resolved.

3.3   Non-attendance by a party

We have considered the non-attendance of a party at the preliminary meeting in Chapter 2 but what if a party refuses to take part in the proceedings at all? This is a situation covered by section 41(4) of the 1996 Act. The first thing that the arbitrator has to do is to ascertain that the party not taking part does not have a good reason for absenting himself from the proceedings. If there is good cause for the failure the arbitrator is not empowered to proceed. Section 41(4) does however allow the arbitrator, provided that due notice has been given and where there is no good cause for the party absenting itself, to proceed in the absence of that party all the way through to his award using, in these circumstances, the evidence presented by one party alone.

The arbitrator must be cautious in these circumstances and, in particular, he must be extremely careful to ensure that he does nothing that could be used by the defaulting party to mount a defence based upon accusations of bias if enforcement of the award is sought in the court.

Other possible situations arise, for example:

- What does the arbitrator do if he is not totally satisfied by the evidence put to him by the one party taking part? Can he award part only of the sum claimed?

    This was the situation in *Fox* v. *P G Wellfair* (1981) 19 BLR 52 which is a salutary lesson to arbitrators, since the arbitrator concerned was highly experienced and qualified. However, he fell into the temptation open to arbitrators of relying on his own expertise in defiance of the evidence. It was an NHBC arbitration with only the claimant represented, since the builder was in liquidation and the NHBC, not being a party to the arbitration, had not put in a defence. The arbitrator heard expert evidence as to defects in support of a claim of £93 000. He then made his own inspection of the defects and, without further reference to either party, made an award in the sum of just over £12 000. The award awas set aside.

    The Court of Appeal ruled:

    o In building contract arbitrations the arbitrator must not form his own opinion and act on his own knowledge without recourse to the evidence.
    o He must receive the evidence of witnesses and the submissions of advocates and reach his conclusions in the light of them.
    o It is not the arbitrator's duty to protect the interests of an unrepresented party by effectively giving expert evidence to himself without giving the represented party the chance to deal with it.

    'If the expert arbitrator, as he may be entitled to do, forms a view of the facts different from that given in the evidence, which might produce a contrary result to that which emerges from the evidence, then he should bring that view to the attention of the parties' (Lord Justice Dunn).

- What happens if he forms the view that there is a reasonable defence to the claim that the absent party, if present, would be bound to use? Say, for example, that case law does not support the claim.

It is suggested that the arbitrator must in such circumstances always ensure that he puts the points in his mind to the party taking part before he uses them in reaching his decision. There might always be a contrary view expressed in another judgment which the arbitrator is not familiar with. One solution may be to send a draft award to the claimant.

3.4 Security for costs

There is no provision in the 1996 Act for the court to deal with an application for security of the respondent's costs in defending the claim (or of the claimant's costs in defending the counterclaim). There is thus no forum other than the arbitrator to deal with such a matter.

This does raise problems both for the parties and the arbitrator. For example a party may wish to reveal the existence of an offer to settle as a part of his submissions on whether or not security should be ordered. This may prove a barrier to a party who may not want the arbitrator to know that an offer has been made, let alone allow him to know the amount of the offer. There is the possibility that a separate arbitrator could be introduced to deal with the question of security alone but this would need the agreement of both parties and the arbitrator and this might not always be forthcoming. Offers to settle are made in almost all arbitrations and they are often made at a level which is higher than that which might ultimately be shown to be the strict contractual entitlement. Any arbitrator should be well aware of the fact that offers are made for purely commercial reasons.

In any event it is incumbent upon an arbitrator to reach his decision on the evidence put before him. Any arbitrator who allows the knowledge of an offer that has been made to affect his ultimate decision is not fit to sit as arbitrator.

The Chartered Institute of Arbitrators has recently prepared 'Guidelines for Arbitrators as to how to approach an application for Security for Costs in terms of section 38 of the Arbitration Act 1996' which are worth referring to when dealing with questions of security for costs. These were published in *Arbitration*, the journal of the Chartered Institute in September 1997.

The detailed mechanics of reaching a decision on whether to award security and if so at what level it should be set are beyond the scope of this book. The arbitrator who is guided by the principles of section 33 of the 1996 Act and allows both parties a reasonable chance to put their respective contentions to him in respect of any application made by a party cannot go far wrong.

Figure 3.1 shows a typical order resulting from an application for security for costs.

FIGURE 3.1 – AN ORDER RELATING TO SECURITY FOR COSTS

In the Matter of the Arbitration Act 1996

and

In the Matter of an Arbitration

between

[*Name of Claimant*]   CLAIMANT

and

[*Name of Respondent*]   RESPONDENT

ORDER No. [   ]
ORDER FOR SECURITY FOR COSTS

The Respondent having made an application to me in writing on [*date*] that the Claimant give Security for costs and the Claimant having replied to that application on [date], I make the following orders:

1. The Claimant [*shall/shall not*] give security for the Respondent's costs of defending the claim.
    Such security shall be given in the sum of £[*amount*].[*Include if appropriate*]

2. The Claimant [*shall/shall not*] give security for my own costs.
    Such security shall be given in the sum of £[*amount*].[*Include if appropriate*]

    [INCLUDE FURTHER POINTS AS FOLLOWS IF SECURITY TO BE GIVEN]

3. This security shall be given by bank guarantee, bond or by payment into a trust account.

4. The Respondent shall pay the Claimant's costs of providing the security if no Order for costs in the Respondent's favour is made.

5. All proceedings in this arbitration shall be stayed pending provision of this security.

6. The costs of this order are reserved.

[*Signed*]
Arbitrator
[*Date*]

The question often arises as to the most convenient way of entitling and enumerating the various orders that are made during the course of the Arbitration. Traditionally orders resulting from procedural meetings have been described as 'Orders for Directions' and orders resulting from applications for security for costs and the like have been headed with the function they perform, 'Peremptory Order' for example. This can mean that you get one series of orders for directions numbered from 1 onwards and perhaps another order or series of orders that may not be numbered at all or also start from 1. This can cause confusion and the authors suggest that the most convenient way is to entitle all orders with the words 'Order No. 1', 'Order No. 2' etc. followed through numerically throughout. If an order performs a specific function that additional function is set out immediately beneath. This is the convention that has been followed in this book.

3.5 Peremptory orders

Arbitrators have always been able to issue peremptory or 'unless' orders to deal with a party's default. Section 41 now formalises the powers of the tribunal in the case of default by a party. This section allows these powers to be agreed between the parties in which case the arbitrator is required to comply with the provisions of that agreement. Alternatively and more likely the parties will not have reached an agreement of this nature and the provisions set out in section 41 to apply in the absence of agreement will come into play.

Section 41(5) allows the arbitrator to make a peremptory order where a party, without sufficient cause, fails to comply with any order or directions of the tribunal. Failure to comply with the peremptory order allows the tribunal to do a number of things. These are:

- In the case where the failure to comply relates to a peremptory order to provide security for costs the arbitrator may dismiss the claim.
- In the case of failure to comply with any other kind of peremptory order the arbitrator has a number of sanctions. These are listed in section 41(7) and are:
  - to direct that the party in default shall not be entitled to rely upon any allegation or material that was the subject matter of the peremptory order;

○ to draw such adverse inferences from the act of non-compliance as the circumstances justify;

○ to proceed to an award on the basis of such materials as have been properly provided to him;

○ to make such order as he thinks fit as to the payment of costs of the arbitration incurred in consequence of the non-compliance.

If all else fails there is always the fall back position (unless the parties have agreed to the contrary) that an application may be made to the court for an order requiring compliance.

The form of a typical peremptory order is set out in Figure 3.2 below.

FIGURE 3.2 – A PEREMPTORY ORDER

In the Matter of the Arbitration Act 1996

and

In the Matter of an Arbitration

between

[Name of Claimant]	CLAIMANT

and

[Name of Respondent]	RESPONDENT

PEREMPTORY ORDER

I made my Order No. [ ] dated *[date]* in which I ordered the following:

> The Claimant is to provide signed copies of the daywork sheets itemised in its claim for payment of Variations Nos 63, 76 and 85.

The Claimant has failed to comply with that Order.

I hereby order that unless the Claimant provides me with the signed daywork sheets that I required to be provided by my Order No. [ ] on or before *[date]* the Claimant shall be debarred from producing the said daywork sheets in evidence and I shall proceed to make my award on the issues to which they relate without regard to them.

The costs of this order shall be paid by the Claimant in any event.

[Signed]
Arbitrator
[Date]

One thing that must be borne in mind is that a peremptory order must be worded identically, as far as the actual directions given, to the order it relates to. The arbitrator must resist the temptation, or the blandishments of a party, to widen the scope of the order. If this were to be done it is doubtful whether the peremptory order would be enforced by the court.

3.6 *Dismissing the claim*

There are two situations where an arbitrator can dismiss a claim, both of which are covered by section 41 of the 1996 Act.

The first is where he is satisfied that there has been inordinate and inexcusable delay on the part of the claimant in pursuing the claim. This is covered by section 41(3). The delay also has to

- either give rise to or be likely to give rise to a substantial risk that it is not possible to have a fair resolution of the issues in that claim, or
- have caused, or be likely to cause, serious prejudice to the respondent.

It should be noted that the delay has to be inordinate *and* inexcusable and at least one of the other two factors must be present before the arbitrator may exercise this power. In general the power should not be exercised unless the statutory period for commencing arbitration has expired since there is otherwise nothing to prevent the claimant from starting the arbitration afresh.

The second instance is where the claimant fails to comply with a peremptory order to provide security for costs. This is covered by section 41(6). The reasoning behind this provision is that it is clearly unfair that an arbitration should continue in existence indefinitely with the associated possibility that the security may be provided at some time in the future and the arbitration thereby resurrected.

Figure 3.3 gives an award dismissing a claim.

FIGURE 3.3 – AN AWARD DISMISSING A CLAIM

In the Matter of the Arbitration Act 1996

and

In the Matter of an Arbitration

between

[*Name of Claimant*]   CLAIMANT

and

[*Name of Respondent*]   RESPONDENT

AWARD

1.00 **Introductory matters**

1.01 By an agreement in writing and under hand dated [*insert date*] in the Form of [*details of contract*] ('the Agreement') the Claimant undertook to carry out for the Respondent as Employer certain works of alteration and repair to the dwelling house known as [*address*].

1.02 Article [*no.*] of the Agreement provides that if any dispute or difference concerning the Agreement shall arise between the Employer or the Architect on his behalf and the Contractor such dispute shall be and is thereby referred to the arbitration and final decision of a person to be agreed between the parties or, failing agreement within 14 days after either party has given to the other a written request to concur in the appointment of an arbitrator, a person to be appointed on request of either party by the President of the Royal Institution of Chartered Surveyors.

1.03 The parties to this arbitration agreement have designated that the juridical seat of this arbitration is England.

1.04 A dispute or difference having arisen, the Claimant applied to the President of the Royal Institution of Chartered Surveyors who appointed me, [*name and qualifications*] of [*address*] to be arbitrator in this reference and I accepted this appointment on [*date*].

1.05 I received an application by the Respondent dated [*date*] that the Claimant give security for the costs of this arbitration.

1.06 The Claimant replied to that application on [*date*].

1.07 I made my Order No. 2 on [*date*] that the Claimant was to give security for the Respondent's costs of defending the claim.

1.08 The Claimant failed to give the ordered security and I made my Peremptory Order on [*date*] to the effect that unless the Claimant provided the security ordered in my Order No. 2 on or before [*date*], I would make an award dismissing the claim.

1.09 The Claimant has failed to provide the ordered security.

1.10 The Respondent has applied to me that I dismiss the Claimant's claim and that I award that all costs incurred by the Respondent shall be paid by the Claimant on an indemnity basis.

2.00 **Award**

2.01 NOW, I [*name of arbitrator*] by reason of the Claimant's failure to provide the security for the Respondent's costs ordered by me hereby dismiss the Claimant's claim in this arbitration.

2.02 The Claimant shall pay the Respondent's costs of the arbitration forthwith on an indemnity basis.

2.03 Should the parties not agree what costs of the arbitration are recoverable I shall determine those costs by award in accordance with section 63 of the Arbitration Act 1996.

2.04 The Claimant shall pay my fees and expenses forthwith which amount to £[*sum*]. Should the Respondent have paid these fees and expenses or any part thereof the Claimant shall reimburse him forthwith.

[*Signed*]
Arbitrator
[*Date*]
Witnessed by ....
Address ....

3.7 *Without prejudice correspondence*

A problem that can sometimes arise is where the arbitrator receives correspondence relating to offers to settle the dispute. The correspondence is privileged from disclosure to the arbitrator and is customarily described as 'without prejudice' correspondence.

The offending correspondence will very often arrive from an unrepresented party in a documents only arbitration. Most arbitrators operate a procedure whereby they have the documents checked by their secretary to identify and remove any privileged documents. These documents are then returned to the party unseen by the arbitrator.

Another method of overcoming this problem is for the arbitrator to leave the submission unopened for a period of a few days, writing to the other party telling him that he is doing so. The other party then has the opportunity to indicate whether there is any correspondence which he considers that the arbitrator should not see. The party submitting the documents should then be given an opportunity to respond before the arbitrator decides whether he should destroy the offending correspondence unread.

The real problem for the arbitrator is when he has inadvertently become aware of the existence of an offer or even of the amount of that offer and one of the parties makes a serious objection to his continuing as arbitrator. This may arise because privileged correspondence does not necessarily have to be marked 'without prejudice' and the arbitrator may inadvertently read such a letter before he realises the nature of its contents. The arbitrator's reaction in this instance must very much depend upon the circumstances. Very often this is a ploy by a reluctant party in an attempt to disrupt the reference. Where the arbitrator considers that this is the real reason behind the objection he should proceed with the arbitration, perhaps indicating the factors that were discussed when considering security for costs earlier in this chapter.

No arbitrator should ever make a decision or refrain from making a decision solely because he is concerned at the consequences to him that might result. For instance an arbitrator should not be deterred from making a decision that he considers right simply because one of the parties threatens to apply to the court for his removal if he does so.

The arbitrator must however examine his position very carefully where he comes into possession of information that he should not really see. On many occasions he will be able to say honestly that his decision will not be influenced. In that event he should proceed. There may however be occasions where the arbitrator feels that he cannot in his heart of hearts continue without the information being likely to affect his decision. In that event he must withdraw.

3.8 Section 34 of the 1996 Act

Section 34, unless the parties have agreed otherwise, provides great opportunities for the arbitrator to introduce procedures which are appropriate to give effect to the principles of sections 1 and 33 of the 1996 Act. This also brings problems in its wake for the unwary arbitrator.

3.8.1 Section 34(2)(a)

For example, this allows the arbitrator to decide where and when the proceedings are to be held. This gives great flexibility but should the arbitrator decide to make the arrangements to suit himself totally, he may find a number of unpleasant side effects. He may find himself liable for the cost of any rooms that he hires if neither party likes his arrangements. He may find himself the subject of an application to the court if he makes arrangements which suit himself alone. It may then be suggested that he has not given each party a reasonable opportunity of putting its case and dealing with that of its opponent, thus flying in the face of the requirements of section 33(1).

3.8.2 Section 34(2)(c)

This allows the arbitrator to decide whether and if so what form of written statements of claim and defence are to be used. This is fairly and squarely aimed at ensuring that statements of case are used rather than pleadings as used in court. This must assist in differentiating between arbitration and litigation and should encourage the settlement of the dispute by enabling each party better to understand the case of the other. The arbitrator may however, if he tries to be too radical, create more problems than he solves. Say, for example, he orders that all that is to be used to identify the case is two final accounts, one produced by the contractor and one by the employer, and in doing so creates a situation whereby neither party can properly put its case or deal with that of its opponent. The arbitrator is immediately in trouble in that he has, by trying to follow section 33(1)(b) too closely, fallen foul of section 33(1)(a).

3.8.3 Section 34(2)(d)

This relates to the disclosure of documents and gives the arbitrator complete discretion, unless the parties have agreed otherwise, to decide which documents or classes of documents should be disclosed by the parties. Where full statements of case are ordered with the parties being asked to provide all the supporting documents with their statements there should be no need for any further documents to be disclosed in the smaller class of dispute. The arbitrator in those cases will be unlikely to have to do more than order the production of the documents with the statements at the outset of the arbitration.

Problems can arise however in more complex disputes. It is unusual nowadays for any arbitrator to be asked to make an order for full discovery and such a request by a party, particularly if it is contested by the other side, should be looked at most carefully before it is granted. Even where the parties are agreed that full discovery should take place the arbitrator should consider the provisions of section 65 of the 1996 Act and think about setting a limit to the recoverable costs either of the arbitration or of the discovery process itself in these circumstances (but not without warning the parties that he is minded to set such a limit and allowing them to make submissions on the point before he does so). It will be far more likely that the arbitrator will receive an application from one party to the effect that documents have not been disclosed by the other party and requesting an order that they be produced within a certain time period. This is a straightforward application for the arbitrator to deal with if it is not contested but it is possible that the party not making the application will object to revealing the documents requested and the arbitrator will have to hear detailed submissions from each party before he decides whether or not to grant the application.

In some extreme cases there may be a question of privilege attaching to the documents and the arbitrator will need to exert a steady hand on the proceedings and possibly take legal advice before making his decision. In this last case where there may be complex legal arguments the arbitrator may be asked to make a reasoned award rather than an order so that the parties have the option of seeking a review by the court of the decision made by the arbitrator. Perceived wisdom is that applications to the court are not generally in the interests of the arbitral process but in this instance, if the matter is of such importance to the parties, the courts will be there to assist and it is far better for the arbitrator to accept that the overall interests of the arbitration are served by making an award rather than insisting on making an order which may end up with one party seeking the arbitrator's removal.

Figure 3.4 is a typical order on an application for disclosure.

FIGURE 3.4 – AN ORDER FOR THE DISCLOSURE OF DOCUMENTS

In the Matter of the Arbitration Act 1996

and

In the Matter of an Arbitration

between

[*Name of Claimant*]	CLAIMANT
and	
[*Name of Respondent*]	RESPONDENT

ORDER No. [   ]

The Claimant applied to me on [*date*] for an Order that the Respondent should give discovery of certain documents. The Respondent responded to the effect that the documents in respect of which discovery was sought were communications between the Respondent and others which were privileged from disclosure to the Claimant and to myself.

I held a hearing lasting 4 hours on [*date*] at which the parties were represented by Counsel.

Having heard and considered the submissions of Counsel for each party I hereby Order that:

the Respondent shall give discovery of all correspondence between itself and [*name*] no later than 7 days from the date of this order.

The costs of the hearing on [*date*] and of this Order shall be paid by the Respondent in any event.

[*Fit for Counsel*]

[*Signed*]
Arbitrator
[*Date*]

3.8.4 Section 34(2)(e)

This allows the arbitrator to decide whether any and if so what questions should be put to and answered by the respective parties and when and in what form this should be done. The arbitrator needs to give careful consideration to the ramifications of this provision before deciding to use it. In many instances this section will provide a useful tool to assist in cases where there are unrepresented parties and the arbitrator decides that it is appropriate for him to act in an inquisitorial manner. This will often prove to be the best way for the arbitrator to ascertain with any certainty what it is that each party wants to get across. Where the parties are represented it might be as well that the arbitrator endeavours to take the parties along with him if he is tempted to dispense with the adversarial procedure that is the norm in this country and operate inquisitorially.

3.8.5 Section 34(2)(f)

Matters relating to evidence are discussed in Chapter 5.

3.8.6 Section 34(2)(g)

The importance of this provision should not be minimised. If not agreed otherwise by the parties the arbitrator has the power to decide whether he should himself take the initiative in ascertaining the facts and the law. Section 34(2)(e) gives the arbitrator a certain latitude to act inquisitorially, section 34(2)(g) increases this latitude. In smaller cases with unrepresented parties this is often the quickest, simplest and most effective way of fulfilling his duty as arbitrator. This is, however, again a provision that could lead the unwary arbitrator into deep water. Before taking this initiative the arbitrator must ensure that he follows the tenets of section 33(1)(b) and that he provides a fair means for the resolution of the matters falling to be determined. The arbitrator will fall foul of this requirement if he decides to ascertain the facts or the law himself and does not give the parties the opportunity to make their own submissions on the material that he ascertains from his investigations.

3.8.7 Section 34(2)(h)

This allows the arbitrator the final word concerning whether there should be oral or written evidence or submissions. Before the 1996 Act came into force if a party insisted on a hearing the arbitrator would be considered to have misconducted himself if he did not allow that hearing. The situation is now different. If the arbitrator considers that an attended hearing is not a procedure that is suitable to the circumstances of the particular case, he is quite at liberty to refuse a party's request that a hearing be held. The arbitrator must however be careful before he does this. He must carefully consider the matters in dispute. Where all the evidence is documentary in nature a hearing is unlikely to do anything to improve that evidence unless the documents themselves need amplifying. Where an arbitrator will get into trouble however is if he shuts the door on an attended hearing and it transpires that there is evidence that can only be assessed properly by putting the witnesses under examination orally. An example is where there are two opposing witnesses who attended a meeting at some stage of the construction works. The minute of that meeting is inadequate and the view of each witness as to the result of the meeting is totally different. One witness will say that there was an agreement on a particular issue and the other says there was not. The only way out of this situation is for the arbitrator to hear these witnesses and decide which one he believes.

3.9   Consolidation of proceedings and concurrent hearings

The power to consolidate proceedings is not one that an arbitrator receives by default. There has to be a positive act on the part of the parties to give this power to the arbitrator which is covered by section 35(2).

It may be that the parties make this decision themselves under section 35(1) in which case all the arbitrator has to do is to proceed in accordance with the agreement between the parties. Alternatively it may be that he is arbitrator on two separate references and the parties may agree to give him the power under section 35(2) to consolidate. In that event the arbitrator will have to satisfy himself that time and costs will be saved by the consolidation. He will have to be sure that all the parties agree, it may be that the same parties are involved in both arbitrations but there is the possibility that the parties are not the same. The arbitrator cannot consolidate one arbitration with another without this agreement.

All the same factors apply to an arbitrator ordering concurrent hearings.

Perusal of the commentary on CIMAR elsewhere in this book will indicate that these rules empower the arbitrator to consolidate but this can only apply if CIMAR apply to both references.

Figure 3.5 sets out an order for consolidation or concurrent hearings.

FIGURE 3.5 – AN ORDER FOR CONSOLIDATING PROCEEDINGS OR ORDERING CONCURRENT HEARINGS

In the Matter of the Arbitration Act 1996

and

In the Matter of Arbitrations

between

[Name of Claimant in dispute No. 1]	CLAIMANT

and

[Name of Respondent in dispute No. 1]	RESPONDENT

and between

[Name of Claimant in dispute No. 2]	CLAIMANT

and

[Name of Respondent in dispute No. 2]	RESPONDENT

ORDER No. [   ]   [*This Order will need to be produced in the same words in each arbitration in the case of concurrent hearings*]

I was appointed as arbitrator in the dispute between [*names of parties in dispute No. 1*] on [*date*] and as arbitrator in the dispute between [names of parties in dispute No 2] on [date].

The parties to these arbitrations have agreed to confer on me the power set out in Section 35 of the Arbitration Act 1996 to [*order the consolidation of the proceedings/to order that concurrent hearings be held*] in both arbitrations.

I hereby ORDER that [*the proceedings shall be consolidated/concurrent hearings shall be held*] in the arbitrations between [*names of parties in dispute No. 1*] and between [*names of parties in dispute No. 2*].

Liberty to apply

The costs of this Order shall be costs in the arbitration[s].

[*Signed*]
Arbitrator
[*Date*]

3.10 Orders for provisional relief

Another power that the arbitrator does not have unless the parties give it to him is the power to make an order for provisional relief (section 39).

This is not, as the side note to section 39 would have one believe, a provisional award.

This power if given relates to any relief that the arbitrator has power to grant in a final award. The phraseology used in the Act is unfortunate in that it leaves in some doubt whether the matter that is dealt with in the provisional order can be dealt with subsequently in anything other than a final award. Common sense must mean that if it is appropriate to make the final decision on a matter dealt with in a provisional order in a partial award under section 47 the arbitrator should do so, but the doubt exists.

It is important to remember that this provision is intended to be used where the arbitrator forms the view that it would not be fair for a party to be kept out of money claimed even though it has not been finally established that he is entitled to it. It is suggested that it would be unusual for the parties to an ongoing arbitration to agree to the powers of the arbitrator being so extended, the one who would have to pay early being likely to object. CIMAR does however include this provision as does the ICE Arbitration Procedure so it is a power that the arbitrator may well be asked to exercise on a regular basis. (*See Figure 3.6.*)

3.11 Arbitration and the decision of an adjudicator

As discussed in Chapter 1 there may be occasions where the parties to a contract with an arbitration agreement will need an arbitrator's award in order to enforce the decision of an adjudicator. The particular concern for the party who has succeeded in obtaining a decision in his favour is how to ensure that the decision is implemented at the earliest possible time especially where there is an active resistance to complying with the decision by the other party.

The adjudicator's decision is binding upon the parties until the dispute is finally resolved by an arbitrator's award, a court judgment or by agreement between the parties. The decision is, however, a matter under the contract and as a result does not have the same status as an arbitrator's award.

Figure 3.6 sets out the form of an order for provisional relief.

FIGURE 3.6 – AN ORDER FOR PROVISIONAL RELIEF

In the Matter of the Arbitration Act 1996

and

In the Matter of an Arbitration

between

[Name of Claimant]   CLAIMANT

and

[Name of Respondent]   RESPONDENT

ORDER No. [   ]
PROVISIONAL ORDER FOR THE PAYMENT OF MONEY

The parties to this arbitration have agreed to confer on me the power set out in Section 39 of the Arbitration Act 1996 to order on a provisional basis any relief which I would have power to grant in a final award.

Having received an application from the Claimant that I order the Respondent to pay certain sums included in the Claimant's final account and having received and considered full documentary submissions from both parties I hereby ORDER that the Respondent shall pay the Claimant the sum of £*[amount]* forthwith.

The costs of this Order shall be costs in the arbitration.

[Signed]
Arbitrator
[Date]

Where there is an arbitration agreement in the contract the court cannot act until it is satisfied that any available arbitral process has been exhausted. One answer is to exclude the enforcement of an adjudicator's decision from the matters that can be arbitrated under the contract and leave it to the court to enforce the decision as a matter of specific performance.

Where, however, there is an arbitration agreement within the contract which does not exclude enforcement of an adjudicator's decisions both the party seeking enforcement and the arbitrator appointed to enforce the decision have to be aware of the possibility that the enforcement proceedings could easily turn into the review of the substantive issues, and thus prevent the adjudication process from working as it should.

It therefore becomes very important from the point of view of the party seeking enforcement that the arbitrator's jurisdiction is limited to the enforcement point only. This means that the notice of arbitration will have to be limited in this way.

It is possible that some parties who have to pay as a result of the adjudicator's decision will seek to put off that payment as long as possible by introducing a set-off or counterclaim. This is where the arbitrator himself comes in. If he is persuaded by the paying party against the wishes of the applicant to widen the dispute and to introduce an investigation of the substantive issues of the dispute into the enforcement proceedings he is effectively preventing the adjudication process from working as it is intended. His award will then have dealt with something not included in the reference to arbitration under which he was appointed and will, as a result, in an all likelihood be subject to appeal.

The practicalities are that the contract between the parties entitles them to a decision which is binding upon them. There is in effect therefore no defence to a reference to arbitration to enforce such a decision. It is not for an arbitrator appointed in these circumstances to be diverted from the task in hand and he should seek to support the adjudication process and make an award enforcing the decision as quickly as he can unless there are exceptionally strong reasons for not doing so.

3.12 Cautionary tales

3.12.1 *Pratt* v. *Swanmore Builders Ltd and Baker* (1980) 15 BLR 37

This case is a warning to all arbitrators. The arbitration arose under a small works contract which provided for an arbitrator to be appointed by the President of the Institute of Arbitrators (as it then was) in default of agreement. The claimant, a Miss Pratt, applied to the Institute for an arbitrator to be appointed, and agreed as a condition of the appointment that the proceedings would be governed by the Institute's procedural rules, to pay the arbitrator's charges and to provide security for those charges if so required. Mr S.T. Baker was appointed and wrote to the parties saying that he accepted the appointment subject to the agreement of his charges. He sent both parties a copy of the Institute's procedural rules but took no steps to get both parties to accept them, nor did he take any steps to ascertain the scope of the agreement or whether it included a term that the Institute's rules should apply. The builders accepted his charges but Miss Pratt did not.

Miss Pratt wrote to the arbitrator, some three months after his initial letter, asking him whether he intended to make an order for a deposit of the whole sum in dispute, plus his charges. The rules did not provide for security for the sum in dispute. Miss Pratt stated that if Mr Baker would make such an order she would make the appropriate deposit and then consider whether to accept his charges. There was further correspondence without progress and a further two months later, Mr Baker called a preliminary meeting at which he ordered that each party should pay £3 500 into a joint account in the names of both solicitors as a deposit in respect of the sum in dispute and his charges.

Miss Pratt paid her share, but the builders did not. Some seven weeks after the order was made the builders' solicitors wrote to Mr Baker challenging his authority to make such an order and said that their clients would be unable to pay the sum required. By this time points of claim and points of defence had been served, but when Mr Baker ordered Miss Pratt to serve further and better particulars of her claim she refused to do so until the builders obeyed his order. A further six months passed before Mr Baker called a further meeting, during which time he made no attempt to ascertain the extent of his powers. At the meeting he ordered each party to pay into court the sum of £500 as security for his charges and a like amount as security for costs in general 'in accordance with [the procedural rules under which] the above arbitration is being conducted as agreed by the parties'.

Two months later – fifteen months since the arbitrator's appointment – the builders offered to put up a bond as security in the sums ordered provided the arbitration could be concluded within a month. Miss Pratt was not sent a copy of the letter; she asked Mr Baker for a copy but he did not send it at that time. Miss Pratt then applied to the Institute under the rules for Mr Baker's removal for failure to use all reasonable dispatch in the conduct of the arbitration. Mr Baker then sent a copy of the builders' letter about the bond to Miss Pratt. Mr Baker refunded the sums she had deposited and gave further directions as to pleadings. The Institute refused to remove Mr Baker and Miss Pratt brought High Court proceedings seeking his removal both on grounds of delay and misconduct.

Her application was successful on the grounds of misconduct. The High Court ruled:

- The arbitrator had misconducted himself and the proceedings and should be removed since there was no prospect that justice would be done to Miss Pratt and indeed the virtual certainty of injustice.
- By allowing himself to be included in the Institute's panel, he had held himself out as a skilled arbitrator and must be held responsible for the conduct of the proceedings.
- The arbitrator had made a fundamental error in failing to establish what the arbitration agreement was and whether the procedural rules had been accepted by both parties.
- He had made other procedural mistakes which went to the root of the arbitration agreement and destroyed Miss Pratt's confidence in his impartiality.

3.12.2 *Bithrey Construction* v. *Edmunds* QBD 29 July 1996; BLISS 1997

In this arbitration the claimants sought to have the arbitrator removed on the grounds of misconduct. The claimants maintained that the arbitrator had conducted the arbitration in such a way that it was apparent to a reasonable man that there was a real danger of bias because:

(a) the arbitrator did not disclose to the claimants that he was acquainted with the respondents' solicitor. The solicitors in question had actually written to the arbitrator and to the claimant advising him to confirm with the claimants that there was no objection to his appointment under the circumstances. The arbitrator did not do so.

(b) the arbitrator considered an application by the respondents that the claimants provide £2 500 as security for costs. This application was resisted by the claimants. The arbitrator sought the advice of his accountant and granted the application without giving the parties the opportunity to consider the material upon which he was proceeding.

The arbitrator was held to have misconducted the proceedings and was removed.

This judgment provides good examples of two of the points that have been considered earlier in this chapter. The arbitrator must not only be sure in his own mind that he will carry out his duties in a totally unbiased way but he must also never put himself into the position of giving the impression to a reasonable man that he might be biased. The arbitrator must also be sure that he tells the parties the results of any investigations that he carries out or has carried out on his behalf.

3.12.3 *Damond Lock Grabowski & Partners* v. *Laing Investments (Bracknell) Ltd* (1992) 60 BLR 112

In this arbitration the arbitrator was appointed to deal with a dispute in respect of a claim for additional monies alleged to be owed by Laing to the contractor, Croudace. Laing then issued proceedings against both Croudace and Damond Lock Grabowski (DLG) claiming against Croudace a declaration that they had no contractual right to any further extension of time or loss and expense and against DLG for negligence/breach of contract and an indemnity in respect of any Croudace claim that might be successful. These proceedings were stayed to arbitration and eventually the same arbitrator was appointed to this arbitration as well.

The reference proceeded through 1991 during which time the arbitrator dealt with various applications and other procedural matters culminating in a hearing in the first arbitration being arranged for December 1991 and a hearing for the second in February 1992. On 1st December 1991 the first arbitration settled with certain immediate consequences the principal of which were an amendment by Laing of their points of claim to reflect the settlement and the disclosure by Laing of some 30 000 documents in January 1992. Laing also served some 66 pages of voluntary particulars which DLG had to deal with at this time.

The arbitrator subsequently made an order at the behest of Laing that the hearing be delayed to early March 1992 but failed to make any reference to DLG before doing so. Upon receiving an application from DLG for an order that the hearing be adjourned so that they might have time to deal with the discovered documents and the voluntary particulars the arbitrator ordered that the hearing date should stand. DLG also requested permission to serve a request for further and better particulars of Laing's amended claim and this was refused. DLG then issued a notice of motion for the arbitrator's removal.

The arbitrator was removed on the basis that he had become obsessed with maintaining the timetable and, objectively judged, it was likely that he could not or would not fairly determine the issues in the arbitration on the basis of the evidence and arguments adduced before him.

Chapter Four
The Hearing

4.1 Introduction

The pre-hearing stage may extend over several months, especially where the issues are complex. The parties will be undertaking their own preparations, in light of the timetable for the hearing set by the arbitrator (see Figure 2.8). In some instances, one party may ask that the hearing be postponed, and this will be dealt with by the arbitrator on its merits as an interlocutory matter.

Assuming, however, that all goes well, the arbitrator must make his own preparations as the hearing date approaches – a fact which should be readily ascertainable from his own diary: A wall-display year planner is very helpful in this connection because hearing dates (and other appointments) have a habit of creeping up by stealth.

Figure 4.1 sets out the basic items, most of which are self-evident.

FIGURE 4.1 – CHECKLIST OF ITEMS NEEDED BY ARBITRATOR AT HEARING

- The File of the Case.
- Notebook(s).
- Pens/ball points/pencils – at least two colours.
- Highlighting pens.
- 'Post-it' stickers.
- Pencil sharpener/ink/refills.
- Watch or small (noiseless!) clock (one with a bleeper can be useful).
- Reference books (e.g., standard books on arbitration and evidence).
- Forms of Oath and Affirmation.
- Appropriate Holy Volumes – New Testament, etc.
- Any documents, references, etc. provided to him before the hearing.
- Calculator: often useful.
- Any personal requirements (e.g. spectacles).

Every arbitration hearing requires some preparation. There are many things which the arbitrator needs to have with him on the day, some common and some less so.

It may be sensible to make an actual list and attach it to one's file for the case involved.

If evidence is to be taken on oath, the arbitrator will need a copy of the New Testament and possibly also the Old Testament. Embarassment can be avoided if the parties are required to bring the appropriate Holy Book, and a translation of the oath if in a language other than English, for any witness who will not swear on the New or Old Testament. He will also need a card setting out the common form of the oath and affirmation. These are obtainable from law stationers or, just as conveniently, may be produced on a word-processor lettered-out on card.

The usual form for the oath in litigation (see the Oaths Act 1978) is as follows:

- For a Christian or Jew –
  I swear by Almighty God that the evidence I shall give shall be the truth, the whole truth and nothing but the truth.

  In arbitration, the form commonly used is –
  I swear by Almighty God that the evidence I shall give touching the matters in difference in this reference shall be the truth, the whole truth and nothing but the truth.

The person taking the oath should hold the Testament uplifted in either hand and say or repeat these words. In Scotland, the form of oath and the manner of its administration are slightly different.

- Those who object to being sworn may solemnly affirm instead, and it is not usual to ask why the person objects to taking the oath.

  The form of affirmation in arbitration to be said or repeated after the arbitrator, is:
  I solemnly, sincerely and truly declare and affirm that the evidence I shall give touching the matters in difference in this reference shall be the truth, the whole truth and nothing but the truth.

The oath is an important matter in the eyes of many arbitrators, witnesses and parties. It adds a certain gravity to the introduction of the witness's evidence and ultimately, if a witness has not told the truth, if it is a serious enough matter that witness can be prosecuted for perjury. Affirming does not let the witness off this particular hook as the same provisions apply in that instance.

Where evidence is for any reason given in writing, the use of an affidavit, which the arbitrator may be empowered to order under applicable rules (CIMAR 5.6 for example) and is in any event empowered to do so by the common law (*Kursell* v. *Timber Operators and Contractors Ltd* [1923] 2 KB 202) gives the evidence the same status as that given under oath.

Much of the arbitrator's time during the hearing will be occupied by taking notes of what is said (see 4.9) and so a good notebook is essential. To some extent the type of notebook is a matter of personal preference, but it should certainly be of an adequate size and a book should not be used to record more than one case. Many arbitrators find the well-known A4 'counsel's notebooks', which have perforations enabling each page to be individually separated, to be the most useful format, while others prefer a hard-bound book or even individual sheets of paper in a ring-binder. In the case of a small arbitration such as an NHBC arbitration, a shorthand notebook may be found more convenient.

Highlighting pens are useful for marking passages in documents according to their importance and the emphasis placed upon them by the parties, and it may be useful to have highlighting pens of several different hues. If pencils are to be used for note-taking, a pencil-sharpener should not be forgotten, nor should ink if a fountain-pen is used. A plentiful supply of writing implements is essential.

If the arbitrator has – as recommended – required the parties to produce photocopies of legal references, there is no need for him to take half the library with him, though he should have access to a good reference library. Certainly, however, he should have available at the hearing one of the standard texts on arbitration for his own reference.

4.2   The timing of the sittings

It is advisable not to start each day's sitting too early or finish it too late. A commonly-used period is from 10 am until 4 pm, with an hour for lunch, which should be taken at some convenient time, for example when there is a break in the evidence.

A five-hour day may not seem very long, but for the arbitrator concentrating on listening to the evidence and taking notes continuously, it is quite long enough. It is also long enough for the others involved, whether as witnesses or advocates. Further, the time before and after the hearing each day (and even during lunch) can be taken up with consultation between the parties and their advisers, or even with negotiations between them, which can be productive of much saving of time overall. If the hearing is a long one – extending, perhaps, over several weeks – everyone concerned (including the arbitrator) will have other things to do. Indeed, in a hearing of more than one week, it is often advisable to sit for only four days a week or to arrange for suitable breaks of a few days at a time.

Allied to the question of timing are matters of creature comfort such as morning coffee, afternoon tea and arrangements for lunch. Many construction arbitrators prefer (wisely) to lunch alone. Coffee and tea are customarily provided, though this may be dispensed with in a formal hearing. Whether one actually breaks the hearing for coffee or tea is a matter of choice and may depend upon the stage the hearing has reached. These sorts of arrangements should have been made by whoever books the room.

4.3   Dress and demeanour

Arbitration is a serious matter. The arbitrator is in a position of great responsibility and should behave accordingly. His demeanour should be formal and such as to command respect.

Those actively and regularly involved in arbitration, whether as arbitrators, advocates or even expert witnesses, inevitably meet socially at conferences and elsewhere and in many cases form personal friendships, as happens between judges and barristers. These personal relationships, however, should never intrude into the hearing, where strict formality should be observed at all times. This extends to the form of address used (e.g. all arbitrators should be addressed as 'Sir' or 'Madam' and, be when the parties are represented, referred to as 'the learned arbitrator').

The formality also extends to dress. A dark suit, a quiet shirt and a subdued tie are the order of the day. Casual dress should certainly be avoided. It is unwise to wear club, old school or professional institution ties or anything indicative of membership of a sodality, since this can lead to embarrassment if, for example, a witness or one of the parties appears wearing the same insignia.

Of course, this concerns the full type of arbitration, akin to a High Court hearing, and less formal dress may be appropriate in smaller arbitrations where the parties are unrepresented. Even here, jeans and a sweater should be avoided and the arbitrator must still conduct himself with reasonable formality. Obviously, the general rules of dress have to be relaxed when inspecting a wet and muddy site, when donkey-jacket, wellington boots and a hard-hat may be the order of the day.

4.4   The day of the hearing

The arbitrator will, of course, ensure that he arrives at the appointed place in good time, taking with him the relevant file and his equipment. He should check that the hearing room is properly laid out and then retire until the parties and their representatives have arrived. In some cases the parties will wish to have their witnesses sitting-in throughout the hearing. The arbitrator should always enquire whether there is any objection from either party to a particular witness or witnesses in general being present and hearing what is said by other witnesses. If objection is made then the witness or witnesses should be directed to leave the room and wait their turn. Since, on occasion, a witness may be called on earlier than expected, any witness excluded in this way should be asked to remain nearby.

Expert witnesses, however, customarily sit-in throughout the hearing. They are there to assist the arbitrator by giving evidence of their expert opinion and this may be affected by the evidence of fact which they hear. They can also often be called upon to assist the arbitrator in relation to the effect of evidence of fact, even before or after they have given their own evidence.

When the appointed hour comes, the arbitrator should enter the hearing room and take his place. He should start by ensuring that everyone who should be there is present and that there is no one present to whom either party has any objection. He then formally opens the hearing. While it is not essential to do so, many arbitrators do this by stating the parties to the arbitration, how he was appointed and that he has accepted the appointment. He should also check that sufficient copies of all the documents have been provided. He should then make a list of all those present by asking them (or the representatives of the parties) to identify themselves or circulate an attendance list. He then invites the claimant to open his case.

In many cases the hearing will start with applications of one kind or another. The following are typical examples:

Amendment of pleadings

Last-minute amendment of pleadings may be necessary because of documents which have only come to the attention of one of the parties when the bundles of documents are being put together just before the hearing.

Challenges to jurisdiction

There may be elements of the case which one or other party considers that the arbitrator has no jurisdiction to deal with. For example, in an arbitration in which there are two respondents, say employer and nominated sub-contractor, there may be a question of a claim between them which it is alleged that the arbitrator cannot deal with since there is no direct arbitration agreement between them.

Applications to call witnesses out of turn

It may be that a witness cannot attend the hearing, for good reason, at the time he would normally be called: e.g. because he is going abroad.

Adjournment of hearing

The arbitrator may be asked, at the outset, to adjourn the hearing for a period, perhaps because someone involved is unavoidably detained elsewhere or negotiations which might lead to a settlement are in progress.

These, and similar problems, are dealt with in Chapter 5. This chapter covers the normal course of events.

4.5 Procedure at the hearing

The normal sequence of events is:

- The claimant opens his case by making an opening statement in which he states the outline as set out in the written statements and gives some indication of the evidence he will call. If there is a counterclaim he will also open his defence to it.

- The claimant calls and examines his witnesses. These may each be cross-examined in turn, in which case they may then be re-examined on matters raised in cross-examination. The arbitrator may also ask questions of a witness if he chooses.
- The respondent may then open his case in a similar way, setting out his defence to the claim and putting forward his counterclaim if there is one.
- The respondent calls his witnesses who are examined, cross-examined and re-examined.
- The respondent sums up his case, reviewing the evidence he has called and the effect which he alleges it to have.
- The claimant does the same.
- The arbitrator closes the hearing stating that he will notify the parties when the award is ready to be taken up.

This sequence of events may change in certain circumstances. Perhaps the commonest is that the respondent says that he does not propose to call any evidence, either oral or documentary. In that case, he may claim the privilege of the last word and the claimant then sums up his case before the respondent. The claimant may challenge this, perhaps on the grounds that the respondent has in fact put forward evidence in the way of documents, etc. during cross-examination of the claimant's own witnesses. The arbitrator must consider such a submission very carefully because, if the claimant is correct, the respondent must lose the right to the last word.

In some cases the whole procedure may be reversed, with the respondent opening his case first. This will be so if the respondent has admitted the claim but wishes to press his counterclaim. In that situation, the counterclaim effectively becomes the claim, and the whole arbitration will be about the counterclaim. A more difficult case in which the respondent may seek to have the procedure reversed is if he alleges that what appears to be his counterclaim is in fact the primary claim because it far outweighs the claim, by amount or otherwise. This can be a very difficult point for the arbitrator to decide and he might be well advised to adjourn the hearing and himself seek legal advice. The arbitrator can of course go a long way towards avoiding the cost of such an adjournment if he holds a pre-hearing review meeting where the parties will have the opportunity to raise problems such as this.

Each of the main elements of the normal procedure are now considered.

4.5.1 Opening statements

Where professional advocates are involved, the opening statements on each side will normally be quite brief. It is the custom nowadays for technical arbitrators to require opening statements to be made in writing and for them to be delivered to the arbitrator in time for him to consider them before the hearing starts. This avoids the (not totally apocryphal) situation of counsel familiarising himself with the case by reading all the documents aloud as his opening statement.

The elements of the case should have been fully set out in the statements of case and defence so that the advocate will only briefly summarise the case and give some indication of the evidence he proposes to call and what he hopes to establish. He may make some comment on the case presented by the other side as well. If there is a counterclaim, the claimant's advocate is likely to set out what he believes to be its basis as it will be put forward by the respondent in order that he may set out his defence to it. If legal points are involved it is useful for the advocate to summarise them and draw the arbitrator's attention to the reported cases and authorities which he proposes to rely on.

Unrepresented parties are sometimes led into making long and irrelevant statements. Whilst it is right that a party must have the opportunity to put his case to the arbitrator as he wishes, the arbitrator must always be guided by section 33(1)(b) and temper this with his duty to avoid unnecessary delay or expense. Written opening statements by unrepresented parties are not always as clear, concise and to the point as the arbitrator might wish and a decision to order these might not have the desired effect. It is perhaps best to accept that opening statements will be given orally in these circumstances and the arbitrator should not be afraid of intervening if the speaker is clearly straying too far from the point or is making abusive statements about the other side. In other words, he should keep control. In some cases he may need to draw out relevant points. Whatever else he does he must avoid the temptation of suggesting to the party what his case ought to be.

4.5.2 Examination of witnesses

As each witness is called, the arbitrator will administer the oath or affirmation as described in 4.1.

The basic rule governing evidence of witnesses is that it should be given orally and in open court so that everyone concerned may hear what is said. This basic rule is often modified both in court and in arbitration. As suggested in Chapter 2 witnesses of fact may be asked to set out their evidence in writing in advance in a statement and the examination would then simply be asking the witness to confirm on oath what he has written and to deal with any points of doubt. Expert witnesses will customarily have set out their evidence in writing and these reports will already have been exchanged between the parties. The essential point is that the other party must have every opportunity of knowing fully what evidence is being advanced against him so that he may answer it.

Unless it has been agreed that witness statements stand as evidence-in-chief the party calling the witness carries out examination-in-chief. Rigid rules apply and the witness must not be asked leading questions. This is a phrase which is much misunderstood and is often used in ordinary conversation to describe a question which is difficult to answer, whereas in fact a leading question is exactly the opposite! A leading question is one which itself suggests its own answer: e.g. 'That instruction was received late, was it not?' It is often said that a question of this sort could be answered 'no', but this is not enough to stop it being a leading question. It is quite clear what answer the questioner expects and that is enough. Questions must be phrased in a neutral form: e.g. 'Did you receive the information?' followed by 'When did you receive it?'

Leading questions may be asked about undisputed matters, such as the name of the witness, his address, qualifications and position. In arbitration, with the agreement of both parties and the arbitrator, leading questions are sometimes admitted, even on disputed facts, in order to shorten the proceedings.

The next stage is cross-examination of the witness by the other party. Its object is to elicit statements about the facts from the witness which are favourable to the other party and also in some cases to discredit the witness so as to establish that he is not a person whose evidence is trustworthy, either because of his character or failings of memory.

Far more latitude is permitted in cross-examination than in examination-in-chief and leading questions may be freely asked. Indeed, they are often the most effective means of extracting from the witness evidence which he really does not wish to give. Questions may be put to the witness dealing with relevant issues not touched on in his examination-in-chief. While a good deal of freedom is allowed at this stage, the questioner should be careful not to go too far, particularly when dealing with the credit or credibility of the witness.

A common trap for the cross-examiner is to introduce evidence in his cross-examination: e.g. by referring the witness to a document for which privilege has been claimed by the questioner's own side. The effect of this would be to make the document admissible and it may reveal matters unfavourable to the questioner.

After cross-examination, the witness may be re-examined by his own side. Re-examination must be confined to matters raised in cross-examination and no new matters may be referred to, although of course it may deal with matters raised in cross-examination which were not in examination-in-chief. Leading questions are not allowed.

4.5.3 Documentary evidence

Documents in any agreed bundle may be referred to without the need for proof, since they will have been agreed between the parties as being genuine and relevant documents – although their exact meaning and significance may be very much in dispute. Documents which are not so included may be introduced in evidence during the hearing, in which case it will be necessary for the party producing the document to prove it.

In general, this will be done by a witness producing the *original* of the document and giving evidence as to its origins: for example by stating on oath that he wrote a particular memorandum. In some cases the other party may challenge whether the document can be admitted in evidence by claiming that it is privileged, i.e. that it is one of the class of documents referred to in 2.6. Where the original document cannot be produced, in some cases a copy may be admissible: for instance where an original letter is in the possession of a third party. It is not permissible for either party to put only part of a document in evidence. If challenged by the other party the entire document must be produced.

4.5.4 Final statements

This is the opportunity for each party to summarise the evidence and show why it favours his case. He should also deal with points of law, whether favourable to him or otherwise, and make his legal submissions as to the effect of any decided cases. Some arbitrators (particularly if they are not legally qualified) ask for any legal submissions to be in writing so that they may consider them at leisure after the hearing, and copies of any such submissions should be provided to the other party.

This can be a very difficult area. Technical arbitrators in general are not expected to be expert lawyers but they must apply the law as it exists, whether they agree with it or not. An arbitrator may consider, on the facts, that one party ought in fairness to win, but the weight of legal submission may be against this. He must follow the law and not his own opinion as to what the outcome ought to be. So, for example, the word 'storm' in ordinary speech is often taken to cover heavy falls of rain, but in the case of *Oddy* v. *Phoenix Assurance Co Ltd* [1966] 1 Lloyd's Rep 134, the word was held to mean some form of violent wind, usually accompanied by rain, hail or snow and did not mean heavy rain as such. An arbitrator in an insurance case would be bound to follow the judgment whatever his private feeling might be as to the merits of the case under the policy in question. Similarly, if dealing with a nominated sub-contractor's delay under the JCT contracts, the arbitrator would be bound to apply the interpretation of the words 'delay on the part of nominated sub-contractors...' laid down by the House of Lords in *City of Westminster* v. *J Jarvis & Sons Ltd* (1970) 7 BLR 64.

The parties' final submissions will need to deal with all matters covered by the claim (or the counterclaim) whether or not they have been the subject of oral evidence. Something that is unlikely to have been dealt with orally is interest upon any amount awarded. It is quite clear from section 49 of the 1996 Act that the award of interest is in the arbitrator's discretion unless the parties have agreed otherwise. The paying party will want to minimise its exposure to interest and the receiving party will want to do the reverse. They will as a result set out in their final submissions the way in which they think the arbitrator should operate his discretion.

Another matter that may need to be included in the final statements is the question of costs. In many arbitrations it is likely that the arbitrator will be unable to deal with costs until he sees the effect that any offer made may have. As a result the arbitrator will need to agree one of two procedures with the parties:

- either to make an award which leaves costs in abeyance until submissions on that aspect are made subsequently, this being the procedure which is currently preferred,
- or to use the 'sealed envelope' procedure and deal with costs on the basis of submissions which he will not read until he has completed that part of his award dealing with the substantive issues.

The former is generally found preferable by parties as they feel that the arbitrator may be unable to resist the temptation of opening the envelope early. It does however have the disadvantage that it costs rather more than dealing with all matters at once and the arbitrator should decide upon the procedure that best suits all the circumstances of a particular case.

4.6   *Rules of evidence*

The arbitrator has the option, by section 34(2)(f), whether or not to apply strict rules of evidence as to the admissibility, relevance or weight of any material sought to be rendered on any matters of fact or opinion. To avoid doubt, this matter is best dealt with before the hearing starts; it is not safe to rely upon any implied right and the parties must know what procedure the arbitrator intends to implement. If evidence is advanced by one party which the arbitrator considers inadmissible he should raise the matter immediately. If the other party agrees to the evidence being adduced, the arbitrator should make a careful note of that agreement and make it clear that he admits the evidence on that basis.

The rules of evidence are many and complex and the arbitrator will need to have ready access to the leading textbooks, such as *Phipson on Evidence*. Where professional advocates are involved the non legally qualified arbitrator will be able to obtain their views if any tricky technical point of evidence arises. Sometimes arguments can arise between the advocates themselves and the arbitrator should require that each point of view is set down in writing so that he may consider them and if necessary seek expert legal advice before he decides whether to admit the evidence or not.

A useful weapon for the arbitrator when faced with this or any other situation where barristers appear to be arguing about matters in which they probably know the answer much better than he does himself is contained in the judgment in *Colt International Limited* v. *Tarmac Construction Limited* CILL April 1996 1145. In that case it was held that it is for the parties' lawyers to assist a non-legally trained arbitrator rather than to put obstacles in his way. A plea to that effect will often have the result of calming down a situation that might otherwise get out of hand.

There are two basic principles:

The evidence must be relevant

Relevance is not a matter of law; it is a matter of common sense and experience. Evidence must be relevant to the facts in issue, although matters outside the strict purview of the arbitration may be referred to if they have some bearing on the points in issue. The evidence must be logically probative of those points. Irrelevant evidence must be excluded.

The evidence must be admissible in law

Whether or not evidence is admissible is a matter of law and the law is largely concerned with what may not be admitted. This is illustrated by the rules relating to evidence of opinion and hearsay evidence.

- Witnesses of fact must testify only as to facts within their own personal knowledge or experience and not as to the inferences to be drawn from those facts. The issues must be determined by the arbitrator, who is entitled to draw inferences, and not by the opinion of witnesses. Some questions inevitably demand an element of opinion in their answer, and this may be admitted in an appropriate case. For instance, a witness's answer to a question about the state of the weather is quite properly admitted, even though it may involve an element of opinion.
- Expert witnesses (see 4.7) may give evidence of their opinions. That, indeed, is their function. It is for the arbitrator to decide whether or not a witness is in fact an expert.
- Hearsay is evidence as to facts expressed through the experience of someone other than the witness and is inadmissible as evidence of the truth of what is asserted. Hearsay evidence may extend to documents and to conduct, but the rule against hearsay was subject to many exceptions, both at common law and by statute, and has to all intents and purposes been abolished by the Civil Evidence Act 1995.

Although the arbitrator has the discretion to allow hearsay statements to be given in evidence, that discretion must be exercised with care, in particular in respect of the weight to be attached to statements put in evidence in this way.

Section 34(2)(f) of the 1996 Act provides that the arbitrator, unless the parties have agreed otherwise, has total discretion with regard to what evidence to admit and what weight to give it if admitted. These powers need to be considered with care. There is a good deal of authority under previous regimes that failure by an arbitrator to give weight to what the court regarded as material evidence amounted to misconduct. It is still arguable that ignoring material evidence is a breach of the requirements of section 33 that the parties must have a reasonable opportunity to put their cases. The arbitrator must be careful to ensure that if he departs from the generally accepted approach to the receiving of evidence, he should be very sure of his ground and probably have taken the precaution of ensuring that the parties are well aware of what he intends to do before he makes his award.

4.7 Expert witnesses

Expert evidence is that given by someone who has the requisite knowledge and experience to form opinions on the basis of facts, whether within his own knowledge or related to him by others. 'An expert may be qualified by skill and experience, as well as by professional qualifications' said Mr Justice Lloyd in *James Longley & Co Ltd* v. *South West Regional Health Authority* (1983) 25 BLR 56. He added:

> 'It is often said that an expert may not be asked the very question which the [arbitrator] has to decide, for the reason that that would be usurp the [arbitrator's] own function. But the dividing line between what an expert witness can and cannot be asked is often very narrow. The ultimate test in all cases of opinion is whether the [arbitrator] needs the evidence in order to reach an informed judgment on the facts.'

A more recent case is *National Justice Compania Naviera SA* v. *Prudential Assurance Co Ltd 'The Ikarian Reefer'* [1993] 2 Lloyd's Rep 68, in which the judgment sets out what is to be expected of an expert in giving his evidence. (*See Figure 4.2.*)

The involvement of experts is a matter customarily included in a preliminary meeting agenda and the number to be allowed on each side and the timetable for their reports is discussed and agreed at that meeting. There are varying views on the method by which the experts should proceed. Some believe that they should exchange preliminary reports before meeting in order to agree facts and figures and to reduce the number of matters in dispute. Others,

---

FIGURE 4.2 – *IKARIAN REEFER* GUIDELINES

1. Expert evidence must be independent and uninfluenced by the exigencies of litigation.

2. It must be unbiased and objective and the witness should not assume the role of advocate.

3. The witness should state the facts or assumptions upon which his opinion is based and should not omit material facts which might detract from his opinion.

4. The witness should make clear whether a particular matter falls outside his expertise.

5. If his opinion is not sufficiently researched due to insufficient data, he must state that his opinion is provisional.

6. If he changes his opinion after exchange of reports, this must be communicated to the other side without delay.

7. Where plans or other documents are referred to in expert evidence they must be supplied to the other party with the report.

---

including the authors, believe that experts should not prepare any report that is discloseable to the other side before the meetings of experts take place. Too often, notwithstanding the fact that an expert is constrained to assist the arbitrator, the preparation of a report that becomes discloseable is likely to harden up the expert's position and diminish the likelihood of agreement.

Expert witnesses are there not primarily to help their clients but to assist the arbitrator to reach a proper conclusion. Where experts conflict in their opinions it will be for the arbitrator to weigh the evidence. Expert evidence may tend to simplify the issues which the arbitrator has to decide and so shorten the proceedings. Conversely, unnecessary expert evidence can prolong the proceedings. Section 37 of the 1996 Act allows the arbitrator to appoint an expert or experts himself who are to report to the arbitrator and to the parties. This will in theory reduce the costs of the arbitration by eliminating duplication. The practical result may mean that there are three experts involved rather than two as the parties may employ their own expert to report on the findings of the arbitrator appointed expert and possibly to give evidence at the hearing to counter the opinions of the arbitrator's appointee. The arbitrator could always exclude the evidence of any expert appointed by a party under section 34(2)(f) but he might well find that he is faced with a

challenge under section 33(1)(a) that he has failed to give a reasonable opportunity to the party to put its case. The appointment of
an expert by the arbitrator is, it is suggested, a course that should
only be undertaken in consultation with the parties, or perhaps as a
result of a decision made where the parties are of opposing views
and one party wants to appoint an expert and the other wants an
appointment to be made under section 37.

4.8   Conduct of the hearing

The arbitrator is in charge of the proceedings but he should not be
intrusive. Essentially, unless he has taken the powers available to
him under section 34(2)(g) to take the initiative in ascertaining the
facts and the law, he is there is listen to the case presented by each
side and not to conduct their cases for them. Nevertheless, he
should not be afraid to intervene if he feels there is a real need for
him to do so if he is to understand the case being presented or if
things are getting out of hand.

Where the parties are unrepresented, it is his duty to keep them to
the point and to ensure that each has a fair hearing. But he must
avoid the pitfall of conducting the arbitration in such a way that he
may appear to be acting unfairly. This is also the case where the
parties are unevenly represented: for example, where one side is
represented by experienced counsel and the other side is representing himself or is represented by an inexperienced lay advocate.
Legal professional etiquette is a great protection in such cases
because counsel will invariably take care not to take unfair advantage of such inexperience and will, indeed, often actively help the
other side in matters of procedure and so forth.

The arbitrator is, of course, entitled to put questions to a witness
himself. However, he should always leave any questions which he
may have until after the witness has been examined by both sides,
unless he thinks that there is a point which ought to be cleared up
immediately in the interests of both parties. Thus, in the normal
case, the arbitrator will ask any questions after the re-examination
of the witness and such questions should not be directed to
obtaining fresh evidence. Their purpose is to clarify any points on
which he is still in doubt.

In the case of a long-winded witness, the arbitrator is entitled to
ask him what he means and to intervene if necessary to keep him to
the point, particularly if he is being questioned by an inexperienced
advocate. Again, however, the arbitrator must refrain from

appearing to take over the case from the advocate (or party) however tempting this course may be. Patience is a great virtue in arbitrators.

Advocates and parties must be allowed to make their opening and closing statements in their own way and the arbitrator should not intervene unless he feels it absolutely necessary to do so. It is often the best course, for instance, at the end of the hearing (if oral closings have been agreed) to allow both sides to make their final statements before asking any questions. The arbitrator's questions at this stage should normally be directed to clarifying points of doubt and he should never be afraid to say that he does not understand a particular point, especially an abstruse legal point, and to ask for further explanations from both sides. He may also ask the advocates to address him on a particular point which he considers of importance and which they have not dealt with in their submissions, especially if it is within his own expertise.

The case of *P.G. Fox* v. *Wellfair* (1981) 19 BLR 52 is worth repeating here.

The Court of Appeal ruled:

- In building contract arbitrations the arbitrator must not form his own opinion and act on his own knowledge without recourse to the evidence.
- He must receive the evidence of witnesses and the submissions of advocates and reach his conclusions in the light of them.
- It is not the arbitrator's duty to protect the interests of an unrepresented party by effectively giving expert evidence to himself without giving the represented party the chance to deal with it.

'If the expert arbitrator, as he may be entitled to do, forms a view of the facts different from that given in the evidence, which might produce a contrary result to that which emerges from the evidence, then he should bring that view to the attention of the parties': Lord Justice Dunn.

At the conclusion of the hearing the arbitrator must be satisfied that he has heard both sides of the case, understood the evidence and submissions and is in possession of all the facts and matters which he needs in order to make his award.

4.9   *The arbitrator's notes*

Note-taking is an art and unless there is a full transcript of the evidence taken down by a shorthand writer, the arbitrator's notes will be the only independent record of the oral evidence. Other than in large complex cases the use of shorthand writers is rare since it is extremely expensive. A system is sometimes used nowadays which utilises computers to give an immediate and direct transcript of the evidence on VDUs for the parties and the arbitrator, who are each able to feed in their own confidential notes which are then provided in hard copy only to them. The arbitrator will however in most cases need to take his own full notes.

Obviously, it is impossible for the arbitrator to record every word spoken *verbatim*, though on occasions it may be necessary for him to take an exact note of the precise words used, in which case he should stop the proceedings to enable himself to do so. Some witnesses speak very quickly, and at the outset of the oral evidence a witness should be asked to keep an eye on the arbitrator's pen or pencil! Experienced advocates will always watch the arbitrator and ensure that he is able to take his notes.

The taking of notes is a useful discipline for the arbitrator, quite apart from its other functions, since it keeps his attention concentrated on what is going on. If the arbitrator is not occupied in this way it is very easy for his attention to wander during long, rambling evidence, however conscientiously he may try to concentrate.

Each arbitrator develops his own method of notetaking and standard abbreviations. There are, however, some essential common points:

- The pages of the notebook should be numbered.
- The notebook should be of the kind that opens sideways and not from the top, so that the right-hand side can be used for the notes of the evidence and submissions and the facing pages can be used for the arbitrator to note comments and points which he may wish to raise himself.
- The time should be recorded occasionally, perhaps at the commencement of each page, and certainly when a witness begins and finishes his evidence, and when breaks are made in the hearing for one reason or another.
- The various stages of the witness's evidence should be clearly indicated. Some arbitrators use different coloured pens for examination-in-chief, cross-examination and re-examination.

120

- Documents adduced during the course of evidence should be separately referenced, for instance by numbering them C/1, C/2, etc for the claimant's documents and R/1, R/2, etc. for those of the respondent.
- A note should be made of the time at which a document is put into evidence.
- Any other exhibits (e.g. photographs, models, etc) should be similarly numbered.
- The name and occupation of each witness should be recorded at the beginning of his evidence and a note made of whether he was sworn or not.
- A very full note should be made of any applications or objections made to the arbitrator and of his decision.

There are various methods of recording questions and answers. The following exchange is typical:

QUESTION: Was a meeting held to discuss these matters?
ANSWER: Yes.
QUESTION: When and where was it held and who was present?
ANSWER: It was held on site on Monday, 4th March. I was there, so was the architect, and I think the consulting engineer was there too, and the QS.
QUESTION: Is there a record of that meeting?
ANSWER: Yes.
QUESTION: This is document 953 in the bundle, is it not?
ANSWER: Yes. That looks like it.
QUESTION: Will you look at the bottom of the last page and read the last paragraph to the learned arbitrator?
ANSWER: It says 'The purpose of the AI is to force their hand and give the consulting engineer the opportunity of terminating the sub-contract on the grounds of non-performance'.
QUESTION: Can you remember who said that?
ANSWER: I don't think those were the precise words used.
QUESTION: They may not be, but can you remember who said something like that?
ANSWER: I think it was me.

This might appear recorded as Figure 4.3.

**FIGURE 4.3 – NOTES MADE BY ARBITRATOR OF THE
EXCHANGE GIVEN IN FULL IN SECTION 4.09**

11.15   Meeting held discuss
     Y

When, where, who
    Site: Mon 4/3. I, arch think c. eng.q.s.

Record
    Y

Document 953
    Y. Looks like

Last page, last para
    Read

Remember who said
    Don't think precise words used

May not, but remember something like
    Think me

        And so on

By starting the witness's response in an indented position rather than halfway across the page it is possible to get far more on one page of notes. Writing constantly over a period of two hours or more at a time can be very tiring, but care should be taken that the notes remain reasonably legible. It may well be a good idea to go over each day's notes after the adjournment, while the evidence is fresh in the mind, and make any clarifying notes, though care must be taken to distinguish those from the notes made at the time. If points of doubt remain, the arbitrator can always ask for clarification at the beginning of the next day's sitting, since if the parties are represented, full notes will have been made on each side.

The arbitrator should record his impressions of the witness, e.g. if he was vague and imprecise, lacking in frankness, etc.

When the arbitrator has made his award, he should still keep his notes in the file. It is the usual practice to keep the file of each arbitration for at least six years. There is always a possibility that those notes may be of importance in some subsequent proceedings and many would advise an even longer period of preservation.

4.10 *Site inspections*

Site inspections are often invaluable, so that the arbitrator may see the subject-matter of the dispute himself. The inspection may take place before, during or after the hearing, and this would be one of the matters decided at the interlocutory stage. If a site inspection has been dispensed with at that stage, one or other party (or the arbitrator himself) may ask for it during the hearing, particularly if it becomes apparent that an inspection will be of assistance.

Before the inspection, the arbitrator should always make it clear whether he will accept evidence from the parties or other witnesses during the visit. It is perhaps more customary for the inspection not to form part of the hearing itself, but for the arbitrator simply to look at the work for himself and for the parties or their representatives to draw his attention to particular matters. The arbitrator should, of course, take a very careful note of what he sees. It may even be advisable to take photographs or arrange for samples, etc. A Polaroid or other instant camera can be of help since the photographs can then be agreed immediately with the parties as a correct record of what the arbitrator has seen.

Chapter Five
Problems at the Hearing

5.1 Introduction

The majority of building contract arbitration hearings proceed quite smoothly, the more so where the parties are professionally represented. However, where lawyers are involved, it is not uncommon for points of law or procedure to be raised, and such problems must be dealt with by the arbitrator.

Other problems can and do arise and the growing body of case law on problem areas adds to the arbitrator's difficulties. If in doubt, the arbitrator should never be afraid to adjourn the hearing to enable him to consider the matter and if necessary to take legal advice. This view should however be tempered with the consideration that adjourning a hearing can be a very costly business and it can often be many weeks or months before it can be reconvened. It may be better to pass on to another matter and deal with the problem the next day after it has been given necessary consideration. It is quite proper for the arbitrator to seek advice from other experienced arbitrators, provided that he maintains confidentiality, for example by not disclosing the names of the parties or the substance of the issues beyond what is necessary to explain the problem. It must always be remembered, however, that the decision eventually given must be that of the arbitrator himself. He is entitled to take other people's opinions into account but must, in the end, make the decision himself in accordance with his own judgement.

This chapter considers some fairly typical problems that can arise during the hearing, in order to give an idea of the many situations which can give rise to difficulties.

5.2 Withdrawal of instructions

Where the parties are legally represented, in principle, matters should be easier for the arbitrator. Legal representation of the parties may, however, in some cases prove disadvantageous from his point of view. A not uncommon instance is where one party or the other withdraws his instructions to his solicitor. The problem might equally arise where a party is represented by a lay advocate. This problem may arise before the hearing; more usually if occurs during the hearing.

For instance, on the morning of the hearing the arbitrator is notified by telephone that one of the parties has withdrawn his instructions from his solicitor and/or advocate. The other party may or may not have been notified, but in any case may object on the grounds that the withdrawal of instructions may cause inordinate delay while new solicitors are instructed.

There is, in fact, very little that the arbitrator can do about this. If the disputant withdraws his instructions from his representatives, he is acting within his rights, and must be given a reasonable opportunity to overcome the problem – which is not necessarily of his making.

Nevertheless, the rights of the other party must also be protected, as he may well not only have been involved in considerable trouble and expense, but – if the withdrawal is during or immediately before the hearing – all the statements will have been exchanged.

Assume the situation where the withdrawal is at the opening of the hearing: the other side may then make an application to the arbitrator to proceed forthwith on the documents, without the need for oral submissions. In such a case, it is suggested that the arbitrator should not accede to the application. The correct course in such circumstances would be for him to adjourn the hearing – say for 28 days – to enable another solicitor to be instructed, but the order (*see Figure 5.1*) should provide that the costs resulting from the adjournment should be borne by the party withdrawing his instructions.

Many permutations of this situation are possible, but the arbitrator's aim must always be to achieve a just result in accordance with the law whilst bearing in mind his obligation under section 33.

One matter that is worth mentioning is the use of the words 'Fit for Counsel'. Before the 1996 Act it was customary for the court to tax (i.e. assess) a party's recoverable costs if they were contested by the other side. This is still the situation in litigation through the courts. In arbitration this has of course been a duty that the arbitrator has been empowered to carry out under the JCT Arbitration Rules 1988 for many years. Under the 1996 Act the assessment of a party's recoverable costs is a matter that may be dealt with by the arbitrator in any situation, and it is only if the

Figure 5.1 is an example of such an order

FIGURE 5.1 – ORDER FOR ADJOURNMENT ON WITHDRAWAL OF INSTRUCTIONS

In the Matter of the Arbitration Act 1996

and

In the Matter of an Arbitration

between

[*Name of Claimant*]   CLAIMANT

and

[*Name of Respondent*]   RESPONDENT

ORDER No. [   ]

Upon hearing Counsel for the Claimant and the Respondent in person regarding the Respondent's application for an adjournment of the hearing upon his withdrawing his instructions from his solicitor and the cross-application on behalf of the Claimant that the arbitration proceed forthwith without a hearing on the basis of documentary evidence and submissions only I now ORDER as follows:

1.   That the hearing be adjourned forthwith and be resumed on [*insert date being not less than 28 days from date of order*] at [*insert time*] and at [*insert place*], this adjournment being for the purpose of enabling the Respondent to instruct a new solicitor.
2.   The Claimant's cross-application is accordingly dismissed.
3.   The costs of the application and of this order and the costs thrown away consequent upon the adjournment be paid by the Respondent in any event.
4.   Liberty to apply.
5.   [Fit for Counsel.]
GIVEN under my hand at [*insert place*] this [*insert date*] day of   19

Arbitrator

---

arbitrator does not do this that the court can get involved. More will be said on this subject in Chapter 6. If the arbitrator is to assess the costs himself the only reason for including the words 'Fit for Counsel' is to assist the parties in understanding what is likely to be allowed by the arbitrator when they seek to agree the costs themselves and perhaps ultimately as an aide memoire for the arbitrator himself.

5.3   *Amendment of pleadings or statements of case*

In the normal course of events, the pleadings or statements of case will have been completed before the actual hearing: see Section 2.5. However, it sometimes happens that, in the process of preparing documents for the hearing, or sometimes as the evidence unfolds, one or other party finds that his case can be better put in an alternative way which he was not aware of when he was preparing his case.

It is clear and settled law that the arbitrator has power to allow the amendment of pleadings even at a late stage. As a matter of principle it must be remembered that the arbitrator is appointed to determine the issues between the parties and it is of paramount importance that those issues should be clearly defined. Even a radical change in a party's case should not be lightly rejected, even though the arbitrator is not bound to allow an amendment: see, for example, *Congimex SARL* v. *Continental Grain Export Corporation* [1979] 2 Lloyd's Rep 396.

The best course in these circumstances is for the arbitrator to make an order allowing the amendment, on terms that the party asking for the amendment should pay the costs involved and further that the other party be allowed to make any consequential amendments to his own pleadings or statement of case.

Each application must be decided on its own merits; the real question is whether justice requires the making of the amendment. Amendments at a very late stage in the proceedings require the most careful consideration. Certainly, where the application is made at the hearing, the arbitrator should consider granting an adjournment. In the case of a lengthy hearing it may be possible to alter the order of the witnesses so that the additional matter can be dealt with as a discrete item at the end of the hearing. If the amendment could have and ought to have been made at a much earlier stage and the amendment sought will cause hardship to the other party, then on balance it should be refused.

127

Figure 5.2 is an example of an appropriate order where amendment is allowed.

Figure 5.3 is an order refusing an amendment.

FIGURE 5.2 – ORDER ALLOWING AMENDMENT OF A STATEMENT OF CASE

In the Matter of the Arbitration Act 1996

and

In the Matter of an Arbitration

between

[*Name of Claimant*]   CLAIMANT

and

[*Name of Respondent*]   RESPONDENT

ORDER No. [   ]

Upon hearing the parties' representatives on both sides and on the application of the Claimant I hereby ORDER and DIRECT as follows:

1. That the Claimant's Statement of Claim be amended in the manner set out in his application.
2. That the Respondent shall, within 14 days of the date hereof, be permitted to make any consequential amendment to the Statement of Defence resulting from the said amendment to the Statement of Claim and shall serve a copy thereof upon the Claimant and upon me within that time.
3. That the hearing in this arbitration stand adjourned until [*insert date, time and place of resumed hearing*].
4. That the Claimant pay the costs of and incidental to the application and this Order and consequent upon the adjournment in any event.
5. Liberty to apply.

GIVEN under my hand at [*insert place*] this [*insert date*]

Arbitrator

**FIGURE 5.3 – ORDER REFUSING AMENDMENT OF
A STATEMENT OF CASE**

In the Matter of the Arbitration Act 1996

and

In the Matter of An Arbitration

between

[*Name of Claimant*]   CLAIMANT

and

[*Name of Respondent*]   RESPONDENT

ORDER No. [   ]

Upon hearing Counsel for both parties and on the application of the
Claimant for leave to amend his Statement of Case, which application
was opposed by the Respondent, I hereby ORDER and DIRECT as
follows:

1. That the Claimant's application be refused.
2. That the Claimant pay the costs of and incidental to the appli-
    cation and this Order in any event.
3. [Fit for Counsel.]

GIVEN under my hand at [*insert place*] this [*insert date*]

Arbitrator

5.4   Adjournment of hearing

It is a fundamental principle that a party to arbitration has a right to
be present throughout the proceedings if he so wishes. He cannot be
excluded without his consent unless, of course, his conduct is so
outrageous that proceedings cannot continue unless he is excluded.
The arbitrator's power to exclude contumacious persons should be
exercised sparingly! Should tempers become frayed – which is not
unknown – the arbitrator's best course is to adjourn the hearing for
a short period to allow them to cool.

It may also be necessary to adjourn the hearing for other reasons.
If, for example, on the first or any day of the hearing, one of the
parties does not appear, and no explanation is offered, the wise
course is to adjourn the hearing to a stated date, ordering the
absentee to pay the costs in the matter. If the absentee is not

represented at the hearing, the arbitrator should write to him appropriately when sending a copy of the order, which should be peremptory.

It is not necessary to write directly to a party who, though represented, is not present personally.

Figures 5.4 and 5.5 are examples of a suggested order and accompanying letter.

FIGURE 5.4 – PEREMPTORY ORDER FOR ADJOURNMENT OF THE HEARING IF PARTY DOES NOT APPEAR

In the Matter of the Arbitration Act 1996

and

In the Matter of an Arbitration

between

[Name of Claimant]   CLAIMANT

and

[Name of Respondent]   RESPONDENT

ORDER No. [ ]

PEREMPTORY ORDER

By my Order No. [ ] dated *[insert date]* I directed that the Hearing in this arbitration be held at *[insert place]* commencing at *[insert time]* on *[insert date]*.

The Claimant was present in person and with Solicitor and Counsel at the appointed date, place and time, but the Respondent failed so to appear either by person or by representatives.

I allowed one hour to elapse from the appointed time before adjourning the Hearing.

NOW I hereby PEREMPTORILY ORDER and DIRECT as follows:

1. That the Hearing in this arbitration stand adjourned and be resumed on *[insert date]* at *[insert time]* at *[insert place]*.
2. If at the date, time and place so appointed the Respondent fails to appear either in person or by representative and continues so to fail for a period of one hour after the time appointed I shall proceed with the Hearing in accordance with section 41(4) of the Arbitration Act 1996 and any Award which I may make following such hearing shall have the same force and effect as if the Respondent had been present throughout the Hearing and had presented his case in full.

---

3. That the Respondent pay the costs thrown away by the adjournment, and the costs of this order, in any event.
4. Liberty to apply.
5. [Fit for Counsel.]

GIVEN under my hand at [*insert place*] this [*insert date*]

Arbitrator

NOTE: *To be sent by recorded delivery or registered post.*

FIGURE 5.5 – LETTER ACCOMPANYING PEREMPTORY ORDER

By Recorded Delivery/Registered Post

To the Claimant

Sir[s]

In the Matter of an Arbitration

[*State names of parties*]

Since you failed to appear either in person or by representative at the date, time and place appointed for the Hearing in accordance with my Order No. [ ] dated [*insert date*] I now enclose my Peremptory Order adjourning the Hearing until [*insert date and time*] at [*insert place*].

You will see from the terms of the enclosed Order that, should you fail to appear again, I shall proceed with the Hearing in your absence.

Yours faithfully,

Arbitrator

A not dissimilar situation arises where the arbitrator is informed that a party or his representative or a material witness is unavoidably detained. Indeed, this problem is quite common where counsel is involved and had been detained on another case, although in that event it is more usual for substitute counsel to make an appearance, even if only to request an adjournment. The best advice is that when in doubt adjourn the hearing, making the appropriate order as to costs.

'There is one panacea which heals every sore in litigation, and that is costs';
Lord Justice Bowen in *Cropper* v. *Smith* (1884) 26 ChD 700.

131

Figure 5.6 is a document which may be adapted as appropriate.

FIGURE 5.6 – ORDER FOR ADJOURNMENT OF THE HEARING IF COUNSEL IS DETAINED ELSEWHERE AND DOES NOT APPEAR

In the Matter of the Arbitration Act 1996

and

In the Matter of an Arbitration

between

[*Name of Claimant*]   CLAIMANT

and

[*Name of Respondent*]   RESPONDENT

ORDER No. [   ]

By my Order for Directions dated [*insert date*] I directed that the Hearing in this arbitration be held at [*place*] commencing at [*insert time*] on [*insert date*].

At the time appointed both Claimant and Respondent were present in person, with their respective solicitors, but only Counsel for the Respondent had appeared.

I was informed by Counsel for the Respondent that a message had been received from the Clerk to Counsel for the Claimant that Counsel had been detained before the Court of Appeal and that no other Counsel was available in Chambers that day.

I ACCORDINGLY ORDER AND DIRECT that the Hearing stand adjourned until tomorrow at the same time and place and I further Order than the question of costs be reserved.

Liberty to apply.

[Fit for Counsel.]

GIVEN under my hand at [*insert place*] this [*insert date*]

Arbitrator

The arbitrator may be asked by those involved to adjourn the hearing. If a joint application is made – for example because the parties are hoping to settle the dispute by negotiation – an adjournment should always be granted, the matter of costs being then reserved. If the negotiations are successful, the arbitrator may then be asked to make an award by consent: see 6.2.4.

5.5 Problems relating to jurisdiction

The question of objections to the arbitrator's substantive jurisdiction at the early stages of the arbitration was dealt with at some length in Chapter 3. Section 31(2) of the 1996 Act widens the opportunity for a party to object allowing such an objection at any stage of the proceedings. Any such objection must be made as soon as possible after the matter that is alleged to be beyond the jurisdiction of the arbitrator is raised. The arbitrator may extend the time for making the objection but only if he considers the delay justified.

It is sometimes the case that an objection is made relating to the arbitrator's jurisdiction in respect of a matter that arises during the hearing. In that event all the matters discussed earlier in this chapter should be considered before making the decision whether to adjourn or continue. It is obvious that an arbitrator must not in any event deal with matters that are outside his jurisdiction.

A factor that the arbitrator will have to keep well aware of is the question of what the contract allows to be arbitrated and at what time. The contract may not allow certain matters to be arbitrated until after practical completion of the works. On the other hand arbitration agreements have in the past allowed the arbitrator to do certain things that might not be expected in the normal run of events, but the 1996 Act now defines the arbitrator's powers so widely that this is unlikely to happen very often in the future.

5.6 Some evidentiary problems

Since the English law of evidence is so complex, it is not surprising that it throws up problems both before and during the hearing. Evidentiary problems can take a variety of forms.

5.6.1 Documentary privilege

In 2.6, something was said about the pre-hearing process of discovery. Questions can, however, arise during the hearing as to whether or not further documents should be produced. A common situation relates to documents for which 'confidentiality' is claimed. The general principle is that *all* documents must be produced – including inter-office memos, 'confidential' records and the like.

The situation is well-illustrated by the case of *Mitchell Construction Kinnear Moodie Group* v. *East Anglia Regional Hospital Board* [1971]

TR 215 where, during an arbitration, the claimant company alleged that the standard of its site personnel was very high. A witness also made reference to annual confidential appraisal reports on each of the company's employees. The respondents contended that site personnel were of low calibre and asked to see the appraisal reports on the personnel concerned.

The claimant refused to disclose the documents and claimed that they were privileged. The arbitrator directed discovery and inspection within 14 days. The claimant refused to disclose the reports and the matter came before the court under the former special case procedure.

Mr Justice Donaldson confirmed the correctness of the arbitrator's order. He said that the test of whether documents should be disclosed was whether they were necessary for *fairly disposing* of the case or for saving costs. If documents were marginally relevant, disclosure should be refused and confidentiality might be a factor which should be taken into account; for example, where A has documents in his possession which concern a trade rival of B, serious issues of confidentiality may arise. Reports as between employer and employees were in a different category. Since arbitration proceedings were private and unreported, if confidentiality was an issue it was only a minor one. In the judge's view the real issue was one of relevance. The test is not whether every document could be used in the case, but whether the documents contain information which may enable a party to advance his case or which may lead to a train of inquiry which has this effect.

On the facts before him, Mr Justice Donaldson said that the issues included:

(1) The effects of the conduct of the architect and contractor; and,
(2) Excessive turnover of the contractor's staff.

He formed the view that these issues indicated fault on the part of the contractor, and, since the documents were capable of allowing a picture of the industrial relations within the contractor's company to be built up, they were relevant, and must be admitted in evidence.

This ruling is of general application and provides guidelines for all arbitrators. Books on evidence discuss the very limited nature of privilege which is effectively confined, in the documentary sense, to communications between a party and his legal advisers.

Minutes of board meetings, pre-tender calculations, diaries, reports, records of telephone conversations and the like are all in principle not privileged.

5.6.2 The ownership of documents

It is sometimes the case that a witness is called at a hearing who is not obviously a part of the party's own organisation, an architect, for example. In the absence of an express provision to the contrary the client who pays for the design service owns the drawings prepared for the project. There should be little problem obtaining the drawings from the architect for the purposes of the arbitration. But what about various documents, correspondence etc. that the architect has generated and received during the course of the project? These are the client's property in accordance with *Beresford (Lady)* v. *Drier* (1852) 22 LJ Ch 407. These documents are however limited to anything that relates to the administration of the contract. Personal memoranda, communications between members of the architect's firm, calculations and private notes remain the architect's property and as such are not within the possession or power of the party to the arbitration and are as a result not discoverable. If however the architect were to be called as a witness in the arbitration and he wished to refer to any of these notes etc. during the course of the hearing, it is submitted that they would have to be made available to the other side prior to the witness giving evidence.

5.6.3 Authenticity of documents

The situation may arise when the parties have reached the hearing without complying with the arbitrator's order to prepare an agreed bundle. Each party then brings the documents to the hearing to which they wish to refer. The arbitrator should adjourn the hearing whilst the parties' documents are consolidated into a single agreed bundle. In this instance even further delay may result if one party does not agree on the authenticity of documents produced by the other. In this event the documents may have to be proved by the party who wishes to use them as evidence.

Similarly even where there is an agreed bundle there may be some dispute as to further documents which a party wishes to produce, the authenticity of which is disputed by the other party. Again those documents would have to be proved and this would have to be done as and when they arise during the course of the hearing.

A technical difficulty arises where one party is not present or represented at the hearing. It is quite likely in that case that the party concerned has refused or been unable to cooperate in the production of the agreed bundle of documents. Technically all documents produced by the party present would then have to be proved. However in practice an arbitrator would only require documents to be proved the authenticity of which could be called into question. For instance, original letters clearly date stamped on receipt could well be accepted as authentic without proof. The case of *Fairclough Building Ltd* v. *Vale of Belvoir Superstore Ltd* [1990] 56 BLR 74 is worth studying on this point.

5.6.3 Oral evidence

Oral evidence is the testimony of witnesses and a number of problems can arise.

The arbitrator cannot himself compel a witness to give evidence. A party to the reference can compel a witness to attend the hearing by applying to the court for a *subpoena ad testificandum*. Such an application can only be made with the permission of the arbitrator or the agreement of the other party (section 43(2)). If following this a witness refuses to attend or give evidence then the party concerned may make application to the court for his committal for contempt of court provided that the writ has been served on the witness not less than four working days before the day upon which he is required to attend the hearing and give evidence, although the court may either shorten or lengthen the four day period.

In most cases, witnesses attend voluntarily without the need for a *subpoena*. Recalcitrant witnesses are not unknown. A witness may attend the hearing and then refuse to give evidence or, more generally, decline to answer a particular line of questioning. If his reason for doing so is that his answers may incriminate him in some way then this is clearly a matter which the arbitrator cannot deal with, and it may then be necessary for the party seeking the evidence to have the whole matter transferred to the court. If the party calling the witness is unrepresented, the arbitrator should suggest that he consider taking legal advice on the question of a *subpoena*.

It often happens that a witness or even one of the parties becomes abusive or interrupts the course of evidence. For example, a party may constantly interject that the witness is telling lies or the witness may become angered by the line of questioning. The arbitrator should administer a sharp warning at the earliest possible stage. If this has no effect, he should adjourn for a short period to allow tempers to cool. If it is one of the parties interjecting, the arbitrator when ordering the adjournment should warn him that if he persists in his conduct after the resumption, he will be excluded from the room while the evidence is being taken, although this is the last resort and should not be done if the party is unrepresented, since he will then be unable to hear the evidence advanced against him. In that case, the arbitrator must be long-suffering and warn the party that his conduct is doing the party no good and simply increasing the costs of the hearing, which will be taken into account in the eventual award of costs. Such a warning is usually effective!

A hearing very seldom passes without one party making some objection to evidence being given by the other party's witnesses. Usually such objections are that the evidence being given is inadmissible: see 4.6. The arbitrator must decide on the validity of the objection and should always ask the other party for his comments on the objection before giving his decision, and if necessary should take a short adjournment in order to think about it quietly.

It may be that the objecting party wants to make a detailed submission on the law relating to the objection. In this event the other side will no doubt wish to respond. If the arbitrator is a technical man he should have made an order at the preliminary meeting that any legal submission should be in writing. There could, as a result, be a considerable delay to the proceedings while this is all sorted out. The problem will be exacerbated if the hearing is drawing to a close and it may be a long time before everyone concerned can be got back together again.

In these circumstances it is quite permissible for the arbitrator to hear the evidence *de bene esse*, that is, provisionally, and to give the parties the opportunity to include their submissions on whether or not he should give any regard to the evidence in question with their closing statements. There may be an objection to this procedure on the basis that once heard, the evidence cannot be forgotten and should it transpire that the arbitrator is persuaded that the evidence should not be taken into account he will have to be very careful to avoid being influenced by it and may indeed need to draw particular attention in his award to this fact.

In general, witnesses of fact should not give evidence of opinion, and an objection on those grounds would usually be sustained. However, in construction arbitrations it is often difficult to draw a clear distinction between the two sorts of evidence. For instance, an architect appearing as a witness of fact may quite properly be asked to state an opinion regarding matters which are within his discretion under the contract, such as quality of work, extensions of time granted and so on. The point is that in such a case it is his opinion which is relevant to the issues: for example, as to the length of extensions of time granted. On the other hand it would not be proper, for instance, for a quantity surveyor to give an opinion as to the quality of workmanship, since that is not within his professional competence, nor is his opinion of any significance under the contract. The quantity surveyor could, of course, state as a fact that he had reduced his valuation of work on the grounds that it was not properly executed, but that would be merely to state his reasons as to why he reduced the valuation. It would not be evidence as to whether in fact the quality was not up to contract standard.

An objection may be made that the facts about which the witness is purporting to give evidence are not within his knowledge. For example, the managing director of a claimant contractor who had never been on site could not give evidence as to what happened on site. If such evidence is given it is hearsay since he would be relying on what someone else had told him and the arbitrator should accordingly give it appropriate weight. Nevertheless, he might properly give evidence to the effect that, having been told of what was happening on site, this was why he took certain actions: e.g. 'As a result of a telephone call from the site agent, I wrote to the architect telling him what I thought'.

5.6.5 Fresh or additional evidence

It sometimes happens that one of the parties may seek the arbitrator's permission to bring in new or additional evidence after he has completed his case. It may be, for instance, that the existence of a document has only just come to light, or evidence given for the other party has made it necessary to bring new evidence in rebuttal. Such applications should always be carefully considered and a decision should never be reached without asking the other party for his submissions. The arbitrator is quite entitled to allow fresh evidence to be brought, though he has a discretion in the matter. The important point is that if the new evidence is really relevant, refusal to admit it may be grounds on which the award could be set aside or remitted by the court. If in the event the evidence proves of no value, the party adducing it can be penalised in costs. If fresh evidence is admitted, the other side must be given the opportunity of rebutting it in turn.

The arbitrator should refuse the application if he is satisfied that the evidence was available at the proper time and ought to have been adduced, though if the applicant is unrepresented some degree of latitude may be allowed for his inexperience in conducting a case.

Should the hearing have finished and new evidence comes to light before the award is made, the arbitrator should, on application, reconvene the hearing to determine the application, and hear the evidence if necessary. For the situation where fresh evidence arises after the award has been made see Chapter 7. Once the award is made the arbitrator only has specific power by section 57(3) to

'(a) correct an award so as to remove any clerical mistake or error arising from an accidental slip or omission or clarify or remove any ambiguity in the award, or

(b) make an additional award in respect of any claim (including a claim for interest or costs) which was presented to the tribunal but was not dealt with in the award.'

In general the arbitrator must be sure not to admit any new submissions or evidence in making such a correction. He must do no more than make any correction he considers to be appropriate in accordance with the submissions already made and evidence already heard.

Chapter Six
The Award

6.1 Introduction

The arbitrator's award is the document by which he conveys his decision to the parties, and like a court judgment, it determines the matters at issue between the parties. The parties are free to agree the form of the award (section 52(1)). In the absence of an agreement otherwise

- the award must be in writing (section 52(3)),
- it must be signed (section 52(3)),
- it must contain reasons (section 52(4)),
- it must state the seat of the arbitration (section 52(5)) and
- it must state the date when it was made (section 52(5)).

The parties could conceivably agree that the award could be oral and certain nineteenth-century cases establish that an oral award will be enforced provided it is communicated to both parties. This is not a course recommended for complicated construction arbitrations! It is not essential for the signature to be witnessed but this is often done.

The award must be:

- *Complete*

The award must deal with and settle all the matters submitted to arbitration, although section 47 of the 1996 Act makes provision for awards on different issues: see 6.2.3 below. This does not necessarily mean that the award must deal separately with every issue in dispute; blanket awards are in fact fairly common in the construction industry. Of course, if the submission to arbitration asks for separate awards, then separate awards must be made.

● *Certain*

The arbitrator's decision must be definite in its terms, so that there is no doubt as to what the parties are required to do. 'It is essential that the document contains an adjudication on the matter in dispute, and not merely an expression of expectation, hope or opinion': Mustill and Boyd, *Commercial Arbitration, Second Edition*, p384. It must also name the parties correctly, and if it does not do so it will not be enforced by the courts: *S G Embiricos Ltd* v. *Tradax Internacional SA* (1967) 1 Lloyd's Rep 464.

● *Consistent*

The award must be consistent within itself. If it is not it will be bad for uncertainty. If the arbitrator's conclusions are clearly inconsistent with his findings or his recitals of the facts the award will be bad.

● *Final*

The arbitrator must deal with all the matters referred to him so that nothing remains to be done. The award must be final on all the issues that it is required to deal with.

6.2   Types of award

Awards are of several types:

6.2.1   Awards for payment of money

These are the most usual type of award in the construction industry. The award directs that one party shall pay to the other a sum of money in full and final settlement of all matters referred to arbitration. The purpose of the award is to compensate the injured party and the sum ordered to be paid is the equivalent of damages at common law.

Figure 6.1 is an example of an award for payment of money.

FIGURE 6.1 – AWARD FOR PAYMENT OF MONEY

In the Matter of the Arbitration Act 1996

and

In the Matter of an Arbitration

between

[*Name of Claimant*]   CLAIMANT

and

[*Name of Respondent*]   RESPONDENT

AWARD

1. This arbitration arises from an agreement in writing and under hand dated [*insert date*] in the Form of Agreement for Minor Building Works issued by the Joint Contracts Tribunal 1980 Edition and incorporating Amendments Nos. 1–10 ('the Agreement') under which the Claimant as Contractor undertook for the Respondent as Employer to carry out certain works of alteration and repair to the dwelling house known as [*insert address*].

2. Article 4 of the Agreement provides that if any dispute or difference concerning the Agreement shall arise between the Employer or the Architect on his behalf and the Contractor such dispute or difference shall be and is thereby referred to the arbitration and final decision of a person to be agreed between the parties or, failing agreement within 14 days after either party has given to the other a written request to concur in the appointment of an arbitrator, a person to be appointed on the request of either party by the President of the Royal Institution of Chartered Surveyors.

3. A dispute or difference having arisen, and the parties having failed to agree upon an arbitrator, the President of the Royal Institution of Chartered Surveyors upon the application of the Claimant has appointed me [*insert name and qualifications*] of [*insert address*] arbitrator and I accepted the appointment by letter to the President dated [*insert date*].

4. The seat of this Arbitration is England.

5. This Arbitration has been conducted under the Construction Industry Model Arbitration Rules [*date*].

6. The parties have agreed that this Arbitration be conducted on the basis of written submissions and documentary evidence and on my inspection without a formal hearing.

142

I NOW, having considered the representations of the parties and the documents submitted by them in evidence and on the basis of my inspection of the additional works **FIND AND AWARD AS FOLLOWS:**

1. There is no dispute as to the extent of the additional work carried out by the Claimant. The dispute is solely regarding the valuation of that work.
2. The Claimant's case is that the work should be paid for wholly on a daywork basis. The Respondent's case is that the work should be measured and valued at rates which have been fixed by the Architect, to which the Claimant responds that if that is so the rates fixed by the Architect are in most cases too low.
3. I agree with the Respondent that dayworks is not the proper basis for the valuation of the additional work and that the work should be measured and valued. However I agree with the Claimant that certain of the rates fixed by the Architect are too low having regard to rates for similar work included in the Claimant's detailed quotation for the original work. **I THEREFORE FIND AND HOLD** that the proper value of the additional work is £3 500.00 and that the Claimant is therefore entitled to a further payment of £1 500.00 plus interest from the date when payment ought to have been made under the Agreement which I find to be [*insert date*] to the date of this Award.

I THEREFORE AWARD AND DIRECT THAT

1. Within 14 days after the date of this Award the Respondent shall pay the Claimant the sum of £1 800.00 (one thousand eight hundred pounds), which sum includes interest under section 49(3) of the Arbitration Act 1996 for the period commencing upon [*insert date*] and ending upon the date of this Award, in full and final settlement of all matters in dispute referred to me in this Arbitration.
2. The Respondent shall pay my fees and expenses in this Arbitration which I determine to be the sum of £[*insert figure and words*] including Value Added Tax and if the Claimant shall have paid the whole or any part of the said costs the Respondent shall reimburse to the Claimant the sum so paid (less any Value Added Tax recoverable by the Claimant).
3. Neither party being legally represented I make no other order as to the costs of this Arbitration.

4. Under section 49(4) of the Arbitration Act 1996 the outstanding amount of this Award including interest and my costs so far as they are reimbursable by one party to the other shall bear interest at 4% above the Midland Bank base lending rate compounded at monthly rests from the date of signature of this Award until payment.

(Signed)
Arbitrator
Date

Witness . . . .

Address . . . .

6.2.2 Awards for specific performance

Although section 48 of the 1996 Act empowers the arbitrator to make an award for specific performance of any contract other than one relating to land this remedy is not appropriate in building contract cases. Specific performance is an order directing that the contract be performed and can only be awarded where damages would be totally inadequate. Courts will not grant specific performance of a building contract (*Hepburn* v. *Leather* (1884) 50 LT 660) and neither should arbitrators, although they are sometimes asked to make, in effect, an award of specific performance, where it is a matter of putting right defective work. As a general rule it is unusual for an arbitrator to be asked to order specific performance. In larger contracts where there are professional advisers involved those advisers and the contractor will invariably resolve questions of defective work between themselves, the only dispute arising being the question of whether the work was in fact defective and who should pay for the remedial works. In smaller contracts where the relationship is between for example a house owner and a builder, by the time that a dispute has reached the stage of arbitration the last thing that the house owner wants is the builder with whom he has fallen out coming back and doing any work at all in his house. There is no possibility of the arbitrator being asked to order specific performance in these circumstances. The house owner has generally had another builder in to remedy the works he alleges to be defective and the argument is again one of money.

It does happen on occasion that circumstances arise where a contractor is refusing to carry out work that has been requested and the relationship between the parties is still good. In this instance, where there is no architect involved, it is quite likely that the arbitrator will be asked by the house owner to order that the builder makes good defective work.

The arbitrator should not, however, direct that the builder put right the defects. What he should do – unless the parties agree otherwise, in which event he is bound by their right to decide the procedure – is to specify those defects which he considers should be put right and give the builder an opportunity to rectify them to the houseowner's satisfaction, without making an award. This should be done by letter, and the houseowner should be asked at the end of the specified period to notify the arbitrator whether he is satisfied or not. If he is, the arbitrator would simply make an award recording the facts and awarding costs against the builder. If dissatisfaction is expressed, the arbitrator would then make an appropriate money award in respect of the defects which have not been properly put right.

Figure 6.2 is an example of the sort of letter referred to.

FIGURE 6.2 – LETTER SPECIFYING DEFECTS TO BE REMEDIED

To the parties   *Date*

Dear sirs,
In the Matter of an Arbitration between
[*Set out names of parties*]

Following my visit to the property on [*insert date*] during which I inspected the defects alleged in the Schedule to the Statement of Claim submitted by the Claimant and heard the representations made to me regarding them by the Claimant's Surveyor and by the Respondent, as agreed I am writing to set out my preliminary findings as to the alleged defects. **THIS IS NOT AN AWARD**.

I find that the following items in the Schedule constitute defects which are due to failure by the Respondent to comply with the contract specification and should be remedied by the Respondent:
Items 1, 3, 4, 6, 7, 10 and 12.
The remaining items do not constitute defects due to failure to comply with the contract specification.

I shall be glad to receive any comments on the above which either party may wish to make within the next 14 days, following which I shall indicate to you within the following 7 days if I then have reason to change my findings as set out above. I suggest that you then make arrangements for the Respondent to carry out the necessary work to remedy those defects within [   ] weeks of the date of this letter. I would ask the Claimant to notify me as soon as the remedial work

---

has been carried out to his and his Surveyor's satisfaction and I will then issue an Award recording the facts and my findings and directing that the Respondent pay my own fees and charges and any other recoverable costs of the Arbitration incurred by the Claimant, including the Surveyor's fees as appropriate.

Yours faithfully,

Arbitrator

At all costs, the arbitrator in these circumstances should avoid putting himself in the situation where he is required to approve the remedial works. Should the remedial works subsequently prove unsatisfactory, there is the possibility that he may be considered to have made a decision under his primary profession and therefore not be protected by the immunity conferred on him by section 29 of the 1996 Act.

6.2.3   Partial awards

Section 47 of the 1996 Act empowers the arbitrator, if he thinks fit, to make more than one award at different times on different aspects of the matters to be determined unless the parties agree otherwise. Before the 1996 Act such an award would have been entitled an interim award but the word 'interim' no longer finds favour. The use of the words 'interim award' was dropped from the legislation because it was seen to cause a misunderstanding as to the status of such an award, with its connotations of a temporary decision. There are two principal situations in which partial awards will be made:

- An award dealing finally with some of the issues, leaving others to be determined later. For instance, the arbitrator might be asked to reserve his award of costs while determining all other issues. His award on the issues will then be a partial award and his eventual award on costs will be the final award.
- An award dealing with a preliminary issue, the decision on which is fundamental to the decision on the other issues. Such awards are often made in respect of liability (i.e. as to whether one party is liable to the other at all) and only if this question is resolved will any question of determining a money award arise. If the party is not liable, this will in effect dispose of the issues.

Partial awards often save time and money and may also be used to decide urgent issues: e.g. liability for defects which require urgent attention, such as a leaking roof. A partial award is just as enforceable as any award that deals with all the issues, provided it meets the other requirements set out in 6.1 above. The first partial award will need to set out the circumstances of the arbitration and the arbitrator's appointment. These will not be repeated in the later awards which will simply refer back to the First award.

One particular situation worth mentioning where a partial award is commonplace under the 1996 Act results from the increased encouragement on the arbitrator to manage the arbitration and instigate procedures which will help to resolve the dispute quickly and effectively. This arises particularly where a dispute relates to both a final account and to loss and expense issues. The arbitrator, particularly where he has experience and knowledge of the costing of the work included in the final account, should, in carrying out his obligations under section 33 of the 1996 Act, instigate procedures whereby he finalises the value of the measured works without the need for full pleadings or statements of case on that particular aspect of the dispute. This will be done by requiring the parties' respective quantity surveyors to meet to agree as much of the final account as they can and identify those items upon which they cannot agree. The arbitrator then meets with the quantity surveyors and makes his declaratory award on the value of the final account. He may also order the payment of monies but he will have to be careful, where there is a substantial counterclaim, to ensure that he takes into account any submissions made by the respondent who is likely to resist the possibility of money changing hands at this time. If the parties have given him the power to make a provisional order under section 39 it may be appropriate to utilise this provision in these circumstances.

It is important to note that although the side note to section 39 suggests that the arbitrator has power to make provisional awards, the text itself refers only to the making of orders on a provisional basis.

Figures 6.3, and 6.4 are examples of partial awards. Figure 6.5 is an example of the recitals to a later award where a partial award has been made.

FIGURE 6.3 – PARTIAL AWARD 1

In the Matter of the Arbitration Act 1996

and

In the Matter of an Arbitration

between

[*Name of Claimant*]   CLAIMANT

and

[*Name of Respondent*]   RESPONDENT

AWARD No. 1

1. This Arbitration arises from an agreement in writing and under hand in the Standard Form of Building Contract Private Edition with Quantities issued by the Joint Contracts Tribunal 1980 Edition incorporating Amendments Nos. 1, 2 and 4 to 15 and dated [*date*] ('the Agreement') under which the Claimant as Contractor undertook for the Respondent as Employer to carry out certain works of construction comprising the erection of factory premises with ancillary works at [*insert address*] and in consideration thereof the Respondent undertook to pay to the Claimant the monies specified in the Agreement in the manner and at the times specified therein.

2. Article 5 of the Agreement provides that if any dispute or difference as to the construction of the Agreement or any matter or thing of whatsoever nature arising thereunder or in connection therewith shall arise between the Employer or the Architect on his behalf and the Contractor either during the progress or after the completion or abandonment of the Works, with certain exceptions not relevant to this Arbitration, it shall be and is thereby referred to arbitration in accordance with clause 41 of the Conditions annexed thereto.

3. A dispute or difference having arisen the parties by agreement have appointed me [*insert name and qualifications*] of [*insert address*] as arbitrator and I accepted the appointment by letter to the Solicitors for the parties dated [*insert date*].

4. The seat of this Arbitration is England.

5. The Arbitration has been conducted under the Construction Industry Model Arbitration Rules [*date*].

6. The parties have agreed that I be asked to make an Award on a preliminary issue as follows:

Whether Architect's Instruction No. 257 constituted an instruction requiring a variation issued under clause 13.2 of the Conditions of the Agreement.

7. A Hearing was held at [*insert location*] on [*insert date*] at which I received certain documentary evidence and heard expert evidence given under oath and the submissions of Counsel for the parties on this issue.

I NOW having considered the evidence and the submissions of Counsel **FIND AND HOLD AS FOLLOWS:**

1. Architect's Instruction No. 257 gives certain details of the entrance canopy which supplement the information set out in Contract Drawing No. 145.

2. The Claimant contends that the details set out in the Instruction involve the execution of work which differs from and/or is additional to that comprised in the definition of the Works under the Agreement and that the Instruction therefore requires a variation within the definition set out in clause 13.1.1 of the Conditions of the Agreement.

3. I find that, while the details set out in the Instruction are additional to those set out in Contract Drawing No. 145, they do not differ from, nor do they involve work additional to that set out in the Contract Bills under the heading of 'Entrance Canopy' on page 238.

I THEREFORE FIND that Architect's Instruction No. 257 did not constitute an instruction requiring a variation issued under clause 13.2 of the Conditions of the Agreement.

AND I AWARD AND DIRECT that my fees and expenses relating to this Award which I determine to be the sum of £[*insert figure and words*] including Value Added Tax shall be paid by the Claimant and if the Respondent shall have paid the whole or any part of such costs the Claimant shall reimburse to the Respondent the sum so paid (less any Value Added Tax recoverable by the Respondent) plus interest of 4% above Midland Bank base lending rate compounded at monthly rests from the date of this Award until payment, and that the Claimant shall also pay the Respondent's recoverable costs of and incidental to the Hearing held on [*insert date*], such costs to be determined by me if not agreed.

FIT FOR COUNSEL

MADE AND SIGNED IN [*insert town*] **ON** [*insert date*]

Arbitrator

Witness . . . .

Address . . . .

FIGURE 6.4 – PARTIAL AWARD 2

In the Matter of the Arbitration Act 1996

and

In the Matter of an Arbitration

between

[*Name of Claimant*]   CLAIMANT

and

[*Name of Respondent*]   RESPONDENT

AWARD No. 1

1. This Arbitration arises from an agreement in writing and under hand in the Standard Form of Building Contract Private Edition with Quantities issued by the Joint Contracts Tribunal 1980 Edition incorporating Amendments nos. 1, 2 and 4 to 15 and dated [*insert date*] ('the Agreement') under which the Claimant as Contractor undertook for the Respondent as Employer to carry out certain works of construction comprising the erection of factory premises with ancillary works at [*insert address*] and in consideration thereof the Respondent undertook to pay to the Claimant the monies specified in the Agreement in the manner and at the times specified therein.
2. Article 5 of the Agreement provides that if any dispute or difference as to the construction of the Agreement or any matter or thing of whatsoever nature arising thereunder or in connection therewith shall arise between the Employer or the Architect on his behalf and the Contractor either during the progress or after the completion or abandonment of the Works, with certain exceptions not relevant to this Arbitration, it shall be and is thereby referred to arbitration in accordance with clause 41 of the Conditions annexed thereto.
3. A dispute or difference having arisen the parties by agreement have appointed me [*insert name and qualifications*] of [*insert address*] arbitrator and I accepted the appointment by letter to the Solicitors for the parties dated [*insert date*], and
4. The seat of this Arbitration is England.
5. This Arbitration has been conducted under the Construction Industry Model Arbitration Rules [*date*].
6. Pursuant to Directions issued by me Statements of Case have been delivered.

7. The parties have agreed that I be asked to make an Award on preliminary issues as follows:
    1. Whether the Claimant is entitled to further extensions of time in addition to those granted by the Architect and if so the additional extension of time to which the Claimant is entitled;
    2. In view of my finding on issue 1, whether the Respondent should repay to the Claimant any part of the liquidated and ascertained damages deducted by the Respondent from payments certified as due to the Claimant and if so to direct the amount to be repaid.
8. A Hearing was held at [*insert location*] on [*insert date*] at which I received documentary evidence and heard evidence of fact and expert evidence given under oath and the submissions of Counsel for the parties on these issues.

I NOW having considered the evidence and the submissions of Counsel **FIND AND HOLD AS FOLLOWS**
1. It is agreed by the parties that there was a period of intense frost during the months of April and May 1996. The Claimant contends that this constituted 'exceptionally adverse weather conditions' within the meaning of clause 25.4.2 of the Conditions of the Agreement and that completion of the Works was delayed thereby for a period of five weeks. The Architect on behalf of the Respondent contends that the weather conditions were not exceptionally adverse having regard to the time of year and the location of the Works; that the Claimant failed to take adequate precautions to protect the Works against the effects of the weather conditions; and that in any case the weather conditions would not have delayed the Claimant had the Claimant been working in accordance with its own programme as submitted at the commencement of the work.
2. **I FIND**
    1. That the weather conditions for the period in question were exceptionally adverse to the progress of the Works having regard to the time of year and the location of the work and therefore fall within the meaning of clause 25.4.2 of the Conditions of the Agreement.
    2. That the Claimant failed to take adequate precautions against the weather conditions and therefore failed to use its best endeavours to prevent delay within the meaning of clause 25.3.4.1 of the Conditions of the Agreement.
    3. That the Claimant's alleged failure to carry out the Works in accordance with its own programme is irrelevant to consideration of whether it is entitled to an extension of time.

4.   That if the Claimant had taken all reasonable precautions available to it to protect the Works against the effects of the weather conditions the completion of the Works would still have been delayed by a period of three weeks as a consequence of those conditions.

I THEREFORE FIND AND AWARD
1.   That the Claimant is entitled to a further extension of time of 3 weeks;
2.   That the Respondent must repay to the Claimant the sum of £1 500.00 (one thousand five hundred pounds), being liquidated and ascertained damages for 3 weeks at £500.00 per week, plus interest compounded at monthly rests from the date 14 days after issue of the Interim Certificate from which the liquidated and ascertained damages were deducted until the date of this award.

I THEREFORE AWARD AND DIRECT AS FOLLOWS:
1.   That the Respondent shall pay to the Claimant the sum of £1 975.00 (one thousand nine hundred and seventy-five pounds) being repayment of liquidated and ascertained damages in the sum of £1 500.00 plus interest under section 49(4) of the Arbitration Act 1996 for the period commencing [*insert date*] and ending on the date of this Award.
2.   That any fees and expenses relating to this Award which I determine to be the sum of £[*insert figure and words*] including Value Added Tax shall be paid by the Respondent and if the Claimant shall have paid the whole or any part of such costs the Respondent shall reimburse to the Claimant the sum so paid (less any Value Added Tax recoverable by the Claimant).
3.   That the Respondent shall pay the Claimant's recoverable costs of and incidental to the hearing held on [*insert date*], such costs to be determined by me if not agreed.
4.   That under section 49(4) of the Arbitration Act 1996 the outstanding amount of this Award including interest and my costs as far as they are reimbursable by one party to the other shall bear interest at 4% above the Midland Bank lending rate compounded at monthly rests from the date of signature of this Award until Payment.

FIT FOR COUNSEL

MADE AND SIGNED IN [INSERT TOWN] ON [INSERT DATE]

Arbitrator

Witness . . . .

Address . . . .

Note: It is not considered appropriate to direct payment of interest under section 49(4) of the Act on a party's recoverable costs until the amount to be paid has been determined.

FIGURE 6.5 – RECITALS TO A FINAL AWARD WHERE A PARTIAL AWARD HAS BEEN MADE

In the Matter of the Arbitration Act 1996

and

In the Matter of an Arbitration

between

[*Name of Claimant*]   CLAIMANT

and

[*Name of Respondent*]   RESPONDENT

FINAL AWARD

1. The circumstances giving rise to this arbitration and to my appointment as Arbitrator are fully set out in my Award No. 1 published on [*insert date*].
2. Following publication of my Award No. 1 and following certain further Directions made by me with the agreement of the parties the parties delivered certain amended Statements.
3. A further hearing on the remaining issues was held at [*insert location*] on [*insert date*] to [*insert date*] at which I received documentary evidence and heard evidence of fact and expert evidence given upon oath and the submissions of Counsel for the Parties

I NOW having considered the documentary and oral evidence and the submissions of Counsel at the Hearing **FIND AND AWARD ON THE REMAINING ISSUES AS FOLLOWS**

[etc. etc.]

6.2.4   Agreed (consent) awards

It is quite common for the parties to settle their dispute before the hearing or before the arbitrator has made his award. In such cases it is usual for the parties to ask the arbitrator to make an agreed award incorporating the terms of the agreed settlement. This procedure is covered by section 51 of the 1996 Act. In any case of settlement the arbitrator should always be asked to approve its terms and he should examine them carefully to make sure that the settlement deals with all the issues in dispute and that the settlement is in terms which are enforceable between the parties.

An agreed award should be in the same form as any other award, but it should make it absolutely clear that the settlement terms are those which the parties have agreed and are therefore not those of the arbitrator's making.

An agreed award is necessary if the 'winning' party's costs are to be determined by the court under section 63(4). Such a reference to the court can only be made if there is an award.

Figure 6.6 is an example of an agreed award.

FIGURE 6.6 – AGREED AWARD

In the Matter of the Arbitration Act 1996

and

In the Matter of an Arbitration

between

[*Name of Claimant*]   CLAIMANT

and

[*Name of Respondent*]   RESPONDENT

FINAL AWARD BY AGREEMENT

1. This Arbitration arises from an agreement in writing and under hand in the Standard Form of Building Contract Private Edition with Quantities issued by the Joint Contracts Tribunal 1980 Edition incorporating Amendments Nos. 1, 2 and 4 to 15 and dated [*insert date*] ('the Agreement') under which the Claimant as Contractor undertook for the Respondent as Employer to carry out certain works of construction comprising the erection of factory premises with ancillary works at [*insert address*] and in consideration thereof the Respondent undertook to pay to the Claimant the monies specified in the Agreement in the manner and at the times specified therein.

2. Article 5 of the Agreement provides that if any dispute or difference as to the construction of the Agreement or any matter or thing of whatsoever nature arising thereunder or in connection therewith shall arise between the Employer or the Architect on his behalf and the Contractor either during the progress or after the completion or abandonment of the Works, with certain exceptions not relevant to this Arbitration, it shall be and is thereby referred to arbitration in accordance with clause 41 of the Conditions annexed thereto.

3. A dispute or difference having arisen the parties by agreement have appointed me [*insert name and qualification*] of [*insert address*] arbitrator and I accepted the appointment by letter to the Solicitors for the parties dated [*insert date*].

---

4. The seat of this Arbitration is England.
5. This Arbitration has been conducted under the Construction Industry Model Arbitration Rules [*date*].
6. The parties have now notified me by letter from the Solicitor for the Claimant dated [*insert date*] that they have reached a settlement of the dispute or difference referred to me in this Arbitration by agreement and have requested that I publish this Agreed Award setting out the terms of that agreement and this has been confirmed by letter from the Solicitor for the Respondent dated [*insert date*].

I NOW in the terms of the agreement reached between the Parties **AWARD AS FOLLOWS**

1. The Respondent shall forthwith pay to the Claimant the sum of £20 000 (twenty thousand pounds) in full and final settlement of all matters referred to me in this Arbitration.
2. The Respondent shall pay my fees and expenses in this Arbitration which I determine to be the sum of £[*insert figure and words*] including Value Added Tax at $17\frac{1}{2}\%$, and
3. The Respondent shall pay the Claimant's recoverable costs of this arbitration, such costs to be determined by me if not agreed.

MADE AND SIGNED IN [INSERT TOWN] ON [INSERT DATE]

Arbitrator

Witness . . . .

Address . . . .

See note to Figure 6.4.

6.2.5   Reasoned awards

Section 52(4) of the 1996 Act requires that the award shall contain the reasons for the award unless it is an agreed award or the parties have agreed to dispense with reasons.

The Departmental Advisory Committee in its Report on the Arbitration Bill made the following comment on the question of the giving of reasons.

'To our minds it is a basic rule of justice that those charged with making a binding decision affecting the rights and obligations of others should (unless those others agree) explain the reasons for making that decision.'

The requirement for giving reasons is therefore not based, as it was in the 1979 Act, on the premise that the only need for reasons is to launch an appeal in the court. The DAC Report goes further, it amplifies the requirement for reasons as follows:

'It was suggested that having to give reasons would be likely to add to the cost of arbitrations and encourage applications for leave to appeal to the Court. We do not agree. The need for reasons is that which we have explained above and has nothing to do with the question of whether or not a Court should hear an appeal from an award.'

Perceived wisdom before the 1996 Act was that the arbitrator should confine his reasons to those which it is essential for a court to have if there is an appeal, that is, he should state only his reasons not his reasoning.

Given the changed emphasis of the 1996 Act and particularly the statements in the DAC report it is submitted that there is now firm encouragement to arbitrators to follow the procedure operated by many construction arbitrators even before the 1996 Act came into force and to set out their reasoning for their findings of fact as well as those of law.

It is worth restating two quotations of Sir John Donaldson, later Lord Donaldson and Master of the Rolls, used in the first edition of this book which, although they were made shortly after the 1979 Act came into force, are still relevant today.

Speaking to the Chartered Institute of Arbitrators, he said:

'I know that some arbitrators feel that giving reasons is a difficult and technical exercise. This is quite wrong. Nothing technical and above all nothing legalistic is required.

The reasons might be as simple as "I award the claimant £500 and costs. I did not believe a word of the evidence called for the respondent. He never delivered the goods. I believe the claimant when he says that he had to buy other goods and that it cost him £500 more than the contract price.'

In his judicial capacity, Sir John Donaldson gave further advice when giving judgment in *Bremer Handelsgesellschaft mbH* v *Westzucker GmbH (No 2)* (1981) 2 Lloyd's Rep 130 CA. He said:

'No particular form of award is required. Certainly, no one wants a formal special case. All that is necessary is that the arbitrators

156

set out what, in their view of the evidence, did or did not happen. They should explain succinctly why, in the light of what happened, they reached their decision and what the decision was. That is all that is meant by a "reasoned award" ...

It is sometimes said that the new approach means arbitrators are delivering judgments and that is something which requires legal skills. That is something of a half truth. Much of the art of giving a judgment lies in telling a story, logically, coherently and accurately. That is something that requires skill, but it is not a legal skill and it is not necessarily advanced by legal training. It is certainly a judicial skill, but arbitrators for the purpose are judges and will have no difficulty in acquiring it.

Where a 1979 Act award differs from a judgment is in the fact that arbitrators will not be expert to analyse the law and the authorities. It will be quite sufficient that they explain how they reached their conclusions. It can be left to others to argue that that is wrong in law and to a professional judge, if leave is given, to analyse the authorities.

That is not to say that where arbitrators are content to set out their reasoning on questions of law in the same way as judges, that will be unwelcome to the courts. Far from it.

The point I want to make is that a "reasoned award" in accordance with the 1979 Act ... is not technical, it is not difficult to draw, and above all, it is something which can and should be produced promptly at the conclusion of the hearing.'

The reference to the 1979 Act in the above quotation does not detract from its relevance.

Reasons should be given in the body of the award.

Figure 6.7 is an example of a reasoned award.

FIGURE 6.7 – REASONED AWARD

In the Matter of the Arbitration Act 1996

and

In the Matter of an Arbitration

between

[*Name of Claimant*]   CLAIMANT

and

[*Name of Respondent*]   RESPONDENT

AWARD

1. This Arbitration arises from an agreement in writing and under hand in the Standard Form of Building Contract Private Edition with Quantities issued by the Joint Contracts Tribunal 1980 Edition incorporating Amendments Nos. 1, 2 and 4 to 15 and dated [date] ('the Agreement') under which the Claimant as Contractor undertook for the Respondent as Employer to carry out certain works of construction comprising the erection of factory premises with ancillary works at [*insert address*] and in consideration thereof the Respondent undertook to pay to the Claimant the monies specified in the Agreement in the manner and at the times specified therein.

2. Article 5 of the Agreement provides that if any dispute or difference as to the construction of the Agreement or any matter or thing of whatsoever nature arising thereunder or in connection therewith shall arise between the Employer or the Architect on his behalf and the Contractor either during the progress or after the completion or abandonment of the Works with certain exceptions not relevant to this Arbitration, it shall be and is thereby referred to arbitration in accordance with clause 41 of the Conditions annexed thereto.

3. A dispute or difference having arisen the parties by agreement have appointed me [*insert name and qualifications*] of [*insert address*] arbitrator and I accepted the appointment by letter to the Solicitors for the parties dated [*insert date*].

4. The seat of this Arbitration is England.

5. This Arbitration has been conducted under the Construction Industry Model Arbitration Rules [*date*].

6. In this Arbitration the Claimant claims
    1. that it is entitled to a further extension of time of three weeks in addition to those granted by the Architect under clause 25 of the Conditions of the Agreement,
    2. the consequential withdrawal of the Architect's certificate issued under clause 24.1 of the Conditions of the Agreement and the issue of a new certificate relating to a Completion Date three weeks later than that fixed by the Architect, and
    3. the repayment by the Employer of the sum of £15 000 being liquidated and ascertained damages for a delay in completion of three weeks at the rate of £5 000 per week stated in the Appendix to the Agreement which the Respondent has withheld from payments certified as due to the Claimant by the Architect.

7. This award is partial in that it does not deal with the determination of the recoverable costs of the arbitration should they not be agreed. Other than in this respect this award deals with all matters referred to me for my decision.

8. A Hearing was held at [*insert location*] on [*insert date*] at which I received documentary evidence and heard evidence of fact and expert evidence given under oath and the submissions of Counsel for the parties.

I NOW having considered the evidence and the submissions of Counsel and for the reasons set out hereafter **FIND AND AWARD THAT THE CLAIMANT'S CLAIM FAILS IN ITS ENTIRETY.**

REASONS
1. The Claimant claims that progress of the Works was delayed by a period of three weeks by a strike of his workpeople on site and that he is therefore entitled to an extension of time of that period by reason of clause 25.4.4 of the Conditions of the Agreement. The Architect on behalf of the Respondent refused to grant any extension of time on the grounds that the Contractor had deliberately induced his workpeople to strike for his own commercial reasons and had made no attempt to settle the dispute reasonably.
2. I find on the facts as shown by the evidence presented to me at the Hearing that the Architect's statement was correct. The Claimant had, for its own commercial reasons, deliberately induced the strike by wrongly withholding overtime payments that were clearly due to its workpeople under Working Rule Agreements.
3. By clause 25.3.4.1 of the Conditions of the Agreement it is a condition of the Contractor's entitlement to an extension of time for any of the Relevant Events listed in clause 25.4 that "the Contractor shall use constantly his best endeavours to prevent delay in the progress of the Works, howsoever caused, and to prevent the completion of the Works being delayed or further delayed beyond the Completion Date". I find from the facts that in this instance the Claimant as Contractor did not use his best endeavours to prevent delay but himself deliberately induced the delay. I therefore find that the Claimant is not entitled to the relief sought in this Arbitration.

I THEREFORE AWARD AND DIRECT AS FOLLOWS:
1. That the Claimant's claim is dismissed.
2. That the Claimant shall pay my fees and expenses in this Arbitration which I determine to be the sum of £[*insert figure and words*] including Value Added Tax and if the Respondent shall have paid the whole or any part of such costs the Claimant shall reimburse to the Respondent the sum so paid (less any Value Added Tax recoverable by the Respondent) plus interest at 4% above Midland Bank base lending rate compounded at monthly rests from the date of this Award until payment.

> 3. That the Claimant shall pay the Respondent's recoverable costs in this Arbitration, such costs to be determined by me if not agreed.
>
> **FIT FOR COUNSEL**
>
> **MADE AND SIGNED IN [*TOWN*] ON [*INSERT DATE*]**
>
> Arbitrator
>
> Witness . . . .
>
> Address . . . .
>
> *See Note to Figure 6.4.*

6.3   Exclusion agreements

By section 69(1) of the 1996 Act the parties may agree to exclude the right of either to appeal to a court on any question of law arising from an award. This is called an 'exclusion agreement'. Previously, under the 1979 Act, such agreements in domestic arbitrations could only be concluded after a dispute had arisen; since section 87 of the 1996 Act was not brought into force they can now be reached at any time including in the arbitration agreement itself.

By section 52(4) of the 1996 Act, the parties may agree to dispense with the requirement for a reasoned award. Since any question of law would only arise from the reasons given in an award, section 69(1) provides that such an agreement has the same effect as an exclusion agreement.

Figure 6.8 is an example of an agreement to dispense with reasons.

FIGURE 6.8 – AGREEMENT TO DISPENSE WITH REASONS

In the Matter of the Arbitration Act 1996

and

In the Matter of an Arbitration

between

[*Name of Claimant*]   CLAIMANT

and

[*Name of Respondent*]   RESPONDENT

AN AGREEMENT made on [*insert date*] between [*insert name and address of Claimant*] and [*insert name and address of Respondent*], who are referred to in this Agreement as 'the parties'.

1. On [*insert date*] the parties entered into a contract in the JCT Standard Form of Management Contract (1987 edition) and in accordance with Article 8 and section 9 of the contract [*insert name of arbitrator*] has been appointed arbitrator to make an award in respect of certain disputes and differences which have arisen between the parties.
2. The parties wish that the arbitrator's decision about the said disputes and differences shall be final.
3. In accordance with section 52(4) of the Arbitration Act 1996, the parties hereby agree that in respect of the arbitration between them the arbitrator shall not give reasons for any award that he makes.

Signed
[Claimant]
[Respondent]

6.4 Form and content of award

There is no set form for an award. Indeed, it has been held that an award which merely says 'I direct that the respondent pay the claimant the sum of £x in relation to all matters referred to me in this arbitration plus costs' is a valid award. Such brevity, however, is not desirable, nor does it comply with the requirement for reasons.

The award is a legal document and must contain enough information to enable the court, if necessary, to enforce it without further inquiry. It must be prepared with the greatest care.

Figure 6.9 is a checklist that may be found useful.

FIGURE 6.9 – CHECKLIST FOR THE AWARD

- **Headings**

— The Arbitration Act 1996
— Name and identify the parties with descriptions: Claimant/Respondent

- **Introductory matter**

— Nature of contract: description, date, etc.
— Refer to arbitration clause
— State method by which appointment to be made, e.g. by agreement of the parties or by appointing authority failing agreement

— State the seat of the arbitration
— State the Rules (if any) under which the arbitration was conducted
— Note that dispute has arisen
— State who has asked for arbitration
— State how you were appointed, e.g. by President of RIBA
— Note acceptance of appointment
— State that directions issued and pleadings served
— State that hearing held or other method, e.g. documents only
— Note having considered oral and documentary evidence and submissions of the parties or representatives as appropriate
— Note whether subject-matter inspected
— State agreement under 1996 Act s52(4) not to require reasons if appropriate
— If a partial award state the issues that it covers

- **The dispute**

— Set out what is claimed and what is counterclaimed
— Set out the list of issues for your decision

- **Findings**

— Set out findings on matters in dispute including reasons

- **Award**

— State 'I now hereby award and direct as follows' or similar wording
— State who is to pay what to whom and when
— Deal with interest
— State in full and final settlement of all matters
— Direct who is to pay costs and on what scale [see 6.6.2 below]
— Deal with interest after award
— State costs to be determined in accordance with section 63(3) if not agreed
— Direct who is to pay arbitrator's costs and expenses
— State amount
— Direct that if payment already made by winning party, losing party to reimburse within specified period
— State fit for Counsel if arbitrator not dealing with assessment of recoverable costs
— State made and published with date and place
— Sign [and witness]

NOTE: This is a checklist for a Final Award which needs review if a Partial Award is being prepared.

Since the parties are probably paying a substantial sum of money, they are entitled to a nice-looking document, properly typed and without manuscript corrections or alterations. A good quality paper should be used; some arbitrators use legal engrossing paper. Others use ordinary heavy paper, such as Conqueror Bond, with a suitable cover and binding. Style and presentation is a matter for individual preference but the advent of word processors and laser printers has made it much easier to produce an impressive-looking document.

Although appearance is important it is the contents of the award which are really essential. The award should always be headed, identifying the applicable Arbitration Act and naming the parties. It is essential that the names of the parties be stated accurately: company reorganisations are not unknown and names may be similar but not exactly the same: e.g. the claimant may start off as Cosdon Contractors Limited, and become transformed into Cosdon Contractors (1998) Limited in the course of the arbitration. The applicable name at the time of the award should be used but the change of name be noted in the recitals, e.g. Cosdon Contractors (1998) Limited (formerly Cosdon Contractors Limited). Occasionally, trading names are used which differ from the official name of the company, and this also should be noted: e.g. John Smith Esq trading as Smith Decorators.

The introductory matters are those parts of the award which set out essential information relating to the arbitration. There should be a reference to the contract itself, again ensuring that it is accurately defined: the date of contract should be correct, and what it related to should be stated. If it is on a standard form, the standard form should be identified precisely, for example as Association of Consultant Architects Form of Building Agreement 1982, 2nd edition 1984. The award should state the juridical seat of the arbitration which has been agreed between the parties or, if so authorised, as directed by the arbitrator himself.

Next there must be reference to the arbitration agreement itself and the fact that a dispute of the kind covered by the agreement has arisen should be stated. The party seeking arbitration should be identified. The method of the arbitrator's appointment should also be stated, whether that appointment was by agreement or made in default of agreement by a third party such as the RIBA President. The fact and date of the arbitrator's acceptance of appointment should be recorded.

This section of the award should then go on to record the form the arbitration proceedings have taken. It should record that

163

directions were issued and pleadings or statements of case were served, if that was the case. If the arbitration has taken some other form, that should be noted, with an appropriate description. The fact that a hearing was held must be given with the dates and location, and the way in which the parties were represented must be stated. If no hearing was held, this should also be stated, with a description of the form the arbitration took instead: e.g. 'with the agreement of the parties the arbitration was conducted on the basis of documentary evidence and written submissions only with no oral hearing'.

If there was a site inspection this fact should be recorded, and reference must be made to the date and who was present at the inspection.

The introductory matters section can, in smaller cases, include a description of the claim and the counterclaim. In larger or more complex cases it is often useful for the award to set out in some detail what it is each party is claiming under a heading of 'The Dispute' or some such. The arbitrator must not fall into the trap of setting out in full each party's pleaded case, this is totally wasteful. The arbitrator should in such a case have required the parties to agree a list of issues that he is to decide upon and this can usefully be set out here.

The next section of the award is the arbitrator's findings of fact. This is essentially a narrative of the events that the arbitrator finds have happened on the basis of the evidence given. In an award which sets out the arbitrator's reasons in full this section could be quite lengthy if there is a multitude of matters to be dealt with in the award. Each individual reason should be kept short and to the point but should be sufficient to enable anyone who reads the award but does not know the circumstances to understand what happened. The award must stand on its own. Even if the parties themselves may not be disputing the facts, it is still desirable that they should be set out, however briefly.

Any findings of law should also be set out, if possible separately from the findings of fact, though in some cases they are so closely intermingled that this is impossible.

Then follow the conclusions drawn from the findings of fact and law and from the reasons. In a lengthy award a heading 'Conclusions' can be useful. In an award dealing with a major final account dispute a section headed 'Summary of Financial Aspects of Award' is also sometimes appropriate.

The operative part of the award then follows and contains the arbitrator's directions as to what is now to happen: that is, who has to pay what to whom, how much and when. The award will have to include a statement regarding interest. This is dealt with more fully in 6.5. The amount of interest awarded and the reasons for arriving at the sum awarded should be given. The arbitrator may decide not to award interest for a part of the period that monies have been outstanding due to delays caused by the 'winning' party himself and in this instance the reasons for this decision should certainly be given. Directions as to payment of costs should be given and there should be a clear statement that the award covers all the matters that it is required to if it is a partial award or all matters in dispute if it is a final award. If the parties were represented by counsel, the award could be endorsed 'Fit for Counsel' by the arbitrator and amounts to a certificate that in the arbitrator's view the engagement of counsel was justified to be shown to the taxing master if the costs are taxed. This is strictly only necessary if the arbitrator is not going to assess the recoverable costs himself.

The award should then state the date and place, and should be signed, and witnessed if the arbitrator considers this appropriate.

It is usual to prepare three copies of the award – one for each of the parties and one for the arbitrator – assuming there are two parties to the dispute.

Section 55 of the 1996 Act, in the absence of agreement otherwise by the parties, requires that the arbitrator sends a copy of his award to each party. This would suggest that the arbitrator keeps the signed original and sends each party a photocopy. Hitherto the general practice has been to send the original signed award to the party who will seek to enforce it with a certified copy being sent to the other party. Alternatively three originals can be produced from the word processor and each be signed in ink. This dispenses with the need to endorse the copies sent to the parties to the effect that they are true copies of the original.

Once the award is completed the arbitrator should notify both parties, stating that it can be taken up by either party on payment of his fees and expenses. Usually it is the claimant who takes up the award. The use of traditional wording such as 'made and published' may still find favour with some arbitrators. Section 55(2) defines the completed award as having been 'made'.

Figure 6.10 is a letter notifying the parties that the award is ready to be taken up.

FIGURE 6.10 – LETTER NOTIFYING THE PARTIES THAT THE AWARD IS READY TO BE TAKEN UP

To The Parties   *Date*

Dear Sirs,

In the Matter of an Arbitration between

[*Set out names of Parties*]

I have now made my Final Award in this arbitration which may be taken up upon payment of my fees and charges in the total sum of £[*insert figure and words*] including Value Added Tax as set out in the attached account.

   Upon receipt of that sum from either party I shall deliver the award to both parties together with my formal receipt to the paying party.

Yours faithfully,

Arbitrator

6.5 *Interest*

There are two types of interest, namely interest which arises as a constituent part of a claim for damages and interest on monies outstanding, i.e. on a debt. The former type of interest is better described as 'financing charges' and is particularly important in construction industry arbitrations because it forms part of the contractor's claim for 'direct loss and/or expense' under JCT contracts (or the equivalent phrase under other contracts).

6.5.1 Financing charges

The principles on which financing charges are based were laid down by the Court of Appeal in *F G Minter Ltd* v. *Welsh Health Technical Services Organisation* (1980) 13 BLR 1 and have been amplified by another decision of the Court of Appeal given in *Rees & Kirby Ltd* v. *Swansea Corporation* (1985) 5 Con LR 34. The essential point established by these decisions is that a building contractor who suffers a loss or expense will have to finance the resulting shortfall in his operations by borrowing money from a bank or by drawing it from money which he would otherwise invest. This financial burden is regarded in law as a constituent part of the debt.

In *Minter*, the contractors were paid amounts in respect of direct loss and/or expense in which they had been involved. They claimed that the amounts so paid were insufficient. Their claim was that, since the amounts in question had not been certified and paid until long after the time when they were involved in the relevant loss or expense, the sum certified ought to have included amounts by way of finance charges for being stood out of their money for long periods. The Court of Appeal accepted this and awarded accordingly.

Under JCT 63 terms – for the contract in *Minter* was substantially in that form – finance charges are recoverable in respect of the period between 'the loss and/or expense being incurred and the making of the written application for reinbursement' by the contractor. Because of a change in wording, successive applications are not necessary under JCT 80 terms, since the relevant clause (clause 26) covers present and future losses, whereas JCT 63 covered only losses already incurred at the time of notice.

Rees & Kirby added a gloss to the *Minter* decision in holding:

- The contractor's application need not be in a specific form but, under JCT 63 terms, 'if it is to be read as relating to financing charges, it must make some reference to the fact that the contractor has suffered loss or expense by reason of being kept out of his money'. Under JCT 80 terms, it may be that there need not be express reference to interest charges.
- The financing charges continue to constitute direct loss or expense until the date of the last application made before the issue of the certificate in respect of the primary loss or expense incurred, e.g. as the result of a variation.
- When the certificate is issued, the contractor's right to receive payment in respect of the primary loss or expense merges in his right to receive payment under the certificate within the time stipulated in the contract. This provides the cut-off point and it seems that the position is the same under JCT 80.
- The interest is real interest, that is interest as paid to a bank on an overdraft or as received on bank deposits, i.e. compound interest, calculated at quarterly or other appropriate intervals, just as a bank operates.

Financing charges are dealt with as part of the award in the arbitration and do not need to be referred to specifically in the directions as to payment.

6.5.2 Interest on the award

The other type of interest, interest on sums awarded, must be dealt with specifically. Historically, the courts have not awarded by way of damages, interest on debts which are paid late: i.e. the courts never recognised that a man is entitled to interest on his money simply because he has been kept out of it for a period of time: *London Chatham and Dover Railway Co* v. *South Eastern Railway Company* [1893] AC 249.

Statute has since affected the position in respect of arbitrators and imposes on them a right and duty to award interest in specified circumstances. The Law Reform (Miscellaneous Provisions) Act 1934 conferred a statutory power to award interest on 'courts of record' and this had been held to extend to arbitrators by implication. The position is now governed by section 49 of the Arbitration Act 1996. The effect of section 49 is that unless a contrary intention is expressed in the arbitration agreement, the arbitrator may award simple or compound interest from such dates, at such rates and with such rests as he considers meets the justice of the case

'(a) on the whole or any part of the amount awarded ..., in respect of any period up to the date of the award;

(b) on the whole or part of any amount claimed in the arbitration and outstanding at the commencement of the arbitral proceedings but paid before the award was made, in respect of any period up to the date of payment.'

The arbitrator may also award simple or compound interest from the date of the award (or any later date) until payment, at such rates and with such rests as he considers meets the justice of the case, on the outstanding amount of any award (including any award of interest included in the award and any award as to costs). Notwithstanding the fact that such an award is at the arbitrator's discretion he should be aware that the award will no longer carry interest automatically as was the case under section 20 of the 1950 Act and he should therefore remember to consider this point when making his award.

Although this is a discretionary power, like all discretionary powers it should be exercised unless there is a good reason for not doing so, and in practice the arbitrator should always award interest for the whole period for which he considers the money to have been outstanding, unless he considers there to be a very good reason to deprive the winning party of this benefit, whether for the whole period or part of it: e.g. he may not award it for that period of time during which he considers the winning party to have been delaying the course of the arbitration. It should be at a commercial rate, which is usually reckoned to be about 2% above base rate. The arbitrator should do this calculation himself and either state the result as a lump sum or simply include it in the overall amount awarded stating that he has done so. To award 'interest at 2% above bank base rate from (date) to (date of this award)' is simply to invite the parties to disagree on the calculation, and with base rates varying between banks and fluctuating as they now do could possibly lead to the award being remitted on the grounds of uncertainty. This comment cannot of course apply to the award of interest on any amounts awarded which remain outstanding from the date of the award. To avoid problems a fixed applicable percentage should be given.

There is nothing to prevent the arbitrator from averaging out the various factors where there are a multitude of individual amounts awarded which become due at different times. The actual base rates at any given time are readily available if enquiry is made of a firm of accountants or a clearing bank.

It is common for construction industry awards to allow a period of grace (e.g. 28 days) for payment, and in such a case interest on the amount outstanding if awarded would be payable from the end of the 28 days.

Figure 6.11 shows an example of an award for interest.

FIGURE 6.11 – THE SECTIONS OF AN AWARD DEALING WITH INTEREST INCLUDING INTEREST TO BE PAID IF PAYMENT OF AWARD IS DELAYED

5.00   Award of Interest

5.01   The Respondent shall pay interest on the amounts claimed in the arbitration and outstanding at the commencement of the arbitral proceedings and paid before this award is made. The interest to be paid in this respect is £[*amount*].

5.02   The Respondent shall pay interest on the amount awarded in respect of the substantive issues in this arbitration in paragraph [*number*] above. The interest to be paid in this respect is £[*amount*].

> 5.03 Should the amounts awarded in this award in respect of the substantive issues in this arbitration in paragraph [*number*] above and in respect of interest in paragraphs 5.01 and 5.02 above remain outstanding beyond the 28 day period set for payment, the Respondent shall pay compound interest on the total amount outstanding, £[*amount*], at x% per annum with three monthly rests from the date of this award until the date of payment.

6.6   Costs

6.6.1   General principles

Under section 61 of the 1996 Act, the arbitrator has discretion to make an award allocating the costs of the arbitration as between the parties. This is subject to any agreement in relation to costs made by the parties. Section 61(2) states the general principle that costs should follow the event, unless the parties agree otherwise, the arbitrator having the discretion to vary this basic principle where he considers it to be inappropriate. The arbitrator must exercise his discretion judicially and according to settled principles and the successful party should not be deprived of his costs unless there are very cogent reasons for so doing.

'The words "judicially exercised" are always somewhat difficult to apply, but they mean that the arbitrator must not act capriciously and must, if he is going to exercise his discretion, show a reason connected with the case and one which the court can see is a proper reason':
   Lord Chief Justice Goddard in *Lewis* v. *Haverfordwest RDC* [1953] 1 WLR 1486.

Mustill and Boyd, *Commercial Arbitration* Second Edition, p 396 lists matters which have been held to justify a departure from the general rule. They include:

- gross exaggeration of the claim;
- unsatisfactory conduct by a party in the course of the arbitration;
- failure by the successful party on issues on which a large amount of time was spent;
- an offer made by the losing party which the other has unreasonably failed to accept (generally where the winning party has

170

failed to accept an offer which was for a greater amount than that ultimately awarded by the arbitrator);
- extravagance in the conduct of the hearing (e.g. employing an excess number of expert witnesses).

Two situations may be considered by way of example. The first is a dispute which involves, say, 15 separate heads of claim. The arbitrator finds in favour of the claimant under five heads, and on the remaining heads makes no award. Should the arbitrator apportion costs?

The answer to this question is that the arbitrator would probably apportion costs to the extent that they were necessarily incurred and no award should be made in respect of costs spent on pursuing obviously invalid claims.

Apportionment supposes, of course, that time was unnecessarily spent on the unsuccessful ten heads. If it were not, then it is suggested that there would be no reason to depart from the general rule.

A second situation is where there is a claim under a single head, and the arbitrator awards one-fifth of the sum claimed. In that case, unless there was a gross exaggeration of the claim, there would be no reason to depart from the general rule. It is not possible to suggest a proportion of award to claim where the arbitrator would or should start to think about apportionment. As Mr Justice McNair observed in *Demolition & Construction Co Ltd* v. *Kent River Board* [1963] 2 Lloyd's Rep 7, there is no principle of law which compels the arbitrator to reflect the measure of success which one party or the other has achieved in his award of costs and it would be dangerous to think otherwise. This matter was also dealt with in *Channel Islands Ferries Ltd* v. *Cenargo Navigation Ltd, 'The Rozel'* [1994] 2 Lloyd's Rep 161.

Problems can arise where there is a counterclaim and much may depend on the actual nature of the counterclaim. If it is in reality a defence, and the claimant succeeds in part, so that the counterclaim also succeeds in part, in general costs should be awarded in favour of the claimant. However, if the counterclaim is a genuine claim, which could have been pursued by separate arbitration if the original claim had not been made, then, if it succeeds in whole or in part, some apportionment of costs should be made in the respondent's favour.

An example of a counterclaim which is essentially a defence would arise where the contractor is claiming disruption and delay and the employer counterclaims for liquidated damages for the equivalent period. Essentially then, the counterclaim is in effect a denial of the contractor's claim to further extensions of time or direct loss and/or expense. If the contractor does not succeed in proving the whole of his claim, then part of the counterclaim must automatically succeed. A counterclaim which is a genuine cross-claim could be in similar circumstances but where the employer counterclaims for the cost of putting right defective work. In that case some apportionment would be appropriate if the counterclaim succeeded in part.

6.6.2 Scales of costs

The method by which the amount of costs which the winning party can recover is determined is set out in section 63(5). The basis used is that there shall be allowed a reasonable amount in respect of all costs reasonably incurred (section 63(5)(a)) and that any doubts whether costs were reasonably incurred or were reasonable in amount are to be resolved in favour of the paying party (section 63(5)(b)). This is what before the 1996 Act would have been referred to as the 'Standard Basis'.

This means that there is generally no need for the arbitrator to make any reference in his award to the way in which these costs are to be determined unless he is of the view that some other basis should be used. In times prior to the 1996 Act it was necessary for the arbitrator to state the basis for the 'taxation' of costs so that, as noted earlier, the taxing master would know the basis upon which he was to work.

The other basis that may be considered by the arbitrator is the 'Indemnity Basis'. This still relates to a reasonable amount in respect of costs reasonably incurred but the yardstick for judging whether costs should be paid is that doubts are resolved in favour of the receiving party. The recovery of costs on this basis would generally be done only when a party has behaved in an unacceptable way. The arbitrator could make an order in respect of costs in this way of his own initiative but it is far more likely that such an order would be made as a result of a specific application by a party to that effect and after the party against whom the order is to be made has had an opportunity to respond before the decision is made.

As far as the award is concerned it would be appropriate for the arbitrator to set out in his award the fact that he considers that costs should be paid on an indemnity basis as there is then guidance for the parties when they come to attempt to agree the amount of the recoverable costs after the award has been made.

6.6.3 The determination of the recoverable costs of the arbitration

Section 63(1) allows the parties to agree what costs of the arbitration are recoverable. The problem arises on many occasions that the paying party considers that the amount claimed by way of costs is far too high. Before the 1996 Act, by section 18 of the 1950 Act there was an implied term in all arbitration agreements that, unless otherwise stated, costs would be taxed by the arbitrator. The parties could choose to override this and have the costs taxed by the court. Under the 1996 Act they have no such option, these costs have to be determined by the arbitrator unless the arbitrator himself declines to do so. It is only in this instance that the parties will have to resort to the court.

The problem for the arbitrator when faced with the requirement to determine the recoverable costs of the arbitration is not the actual doing of the determination. That is relatively easy, he decides what are reasonable costs and whether they have been reasonably incurred and tells the parties the amount to be paid. The problem that he may well be faced with is the expectations of the parties particularly when they are legally represented. Then they may expect the arbitrator to make his determination in accordance with the principles that a taxing master would use, that is in accordance with the Rules of the Supreme Court, Order 62 in particular, and also with legal precedent. There is no way that it is reasonable for the parties to expect a technical arbitrator, however experienced, to have the knowledge of a taxing master who deals with these matters full time. It is for the parties to make the arbitrator aware of any particular legal precedent they expect him to take account of.

It must also be noted that section 63(3) requires that the arbitrator makes his determination by award. By definition a reasoned award is therefore required. Section 63(3) requires that the arbitrator states in his award the basis upon which he has acted and the items of recoverable costs and the amount referable to each. The extent of reasons that may be required by the court if a party seeks to appeal an award determining the recoverable costs has not yet been tested. The award of costs is to a certain extent discretionary, that is, it is for the arbitrator to decide upon the reasonableness of the costs claimed. It is also expected that the arbitrator will have doubts as is intimated in section 63(5)(b). It would in our view be inappropriate to expect more in the way of reasons in an award than a statement regarding the basis upon which the arbitrator has acted, (section 63(3)(a)), the items of recoverable costs and the amount referable to each, (section 63(3)(b)) and a statement that he has made his determination of the costs recoverable in accordance with section 65(5), that is, they are a reasonable amount in respect of costs reasonably incurred.

If the arbitrator should decline to make a determination of the recoverable costs of the arbitration the parties then would have to go to the court if they are unable to agree the amount of the recoverable costs. In that instance for the award to be complete it would be even more necessary for the arbitrator to indicate if he had decided that indemnity costs were appropriate. This might also be a case for using the term 'Fit for Counsel' for the information of the taxing master if the arbitrator considers that the use of counsel has been justified.

Figure 6.12 sets out an example of an award following an application to determine costs.

FIGURE 6.12 – AN AWARD DETERMINING THE RECOVERABLE COSTS OF THE ARBITRATION

In the Matter of the Arbitration Act 1996

and

In the Matter of an Arbitration

between

[Name of Claimant]	CLAIMANT

and

[Name of Respondent]	RESPONDENT

AWARD No. 3

1.00 **Introductory matters**

1.01 The circumstances giving rise to this arbitration and to my appointment as arbitrator are fully set out in my Award No. 1 made on [*date*]

1.02 My Award No. 1 was a Partial Award which dealt with all issues in this arbitration save for the liability for the costs of this arbitration.

1.03 My Award No. 2 dealt with the liability for the costs of this arbitration.

1.04 This Award No. 3 arises as a result of an application by the Respondent dated [*date*] that I determine the recoverable costs of the arbitration.

1.05 The Claimant submitted a detailed note of the costs it seeks to recover on [*date*].

1.06 The Respondent submitted a written schedule of objections to the Claimant's note of costs on [*date*].

1.07 The Claimant submitted a written reply to the Respondent's schedule of objections on [*date*].

1.08 I have considered the submissions of the parties and I have determined the recoverable costs of the arbitration on the basis that there shall be allowed a reasonable amount in respect of all costs reasonably incurred and I have resolved any doubts as to whether costs were reasonably incurred or were reasonable in amount in favour of the paying party.

1.09 My findings in respect of each item claimed by the Claimant are set out in manuscript upon the copy of the Claimant's note of costs appended to this award.

1.10 The Claimant has submitted that the Respondent should pay the Claimant's costs of this determination and award and the Respondent has accepted that this is the customary method of awarding costs in a taxation in the court and is therefore applicable in this arbitration. [This decision might be reversed if the arbitrator considers it appropriate.]

2.00 AWARD

2.01 NOW I, [*name of arbitrator*] determine that the Claimant's recoverable costs in this arbitration including the costs of this determination shall be £[*sum*].

2.02 The Respondent shall pay my fees and expenses relating to this award which I determine in the sum of £[*sum*] including Value Added Tax. If the Claimant shall have paid the whole or any part of this sum the Respondent shall forthwith reimburse the Claimant the amount so paid.

[*signed*]
Arbitrator
[*date*]

Witness . . . .

Address . . . .

6.6.4 The arbitrator's fees

The arbitrator's safeguard for the recovery of his fees is his right to retain his award until his fees are paid. In legal terms he has a lien on the award. This is ratified by section 56(1) of the 1996 Act. However, section 56(2) of the 1996 Act provides that if the arbitrator refuses to deliver his award except on payment of his fees, the court can intervene on application by either party, and order that the award be delivered on payment into court of the fees demanded. Those fees will then be taxed by a taxing master and the arbitrator will be paid the amount the master finds justified and the balance will be paid back to the applicant.

6.6.5 Limiting recoverable costs

One matter that an arbitrator should have considered in the course of the arbitration is the question raised by section 65 of the 1996 Act, that is, the limitation of the recoverable costs of the arbitration or any part of the arbitral proceedings to a specified amount. This is a matter that the arbitrator will have to consider sufficiently before the costs are incurred for the limit to be taken into account.

Often this is a matter that the arbitrator may raise at the preliminary meeting and seek the parties' comments. At the preliminary meeting it is common that the arbitrator can judge from the attitudes of the parties whether it will be necessary to consider imposing such a limit. If the parties are receptive to the suggestions of the arbitrator made with the intention of complying with his obligations under section 33(b) to adopt procedures suitable to the circumstances of the particular case, avoiding unnecessary delay and expense, it may well be that procedures are adopted that are cost effective and unlikely to result in a unnecessarily high expenditure on costs. The arbitrator might then do no more than warn the parties that he will review the position if he considers that they appear at any stage likely to start incurring costs unnecessarily.

If however the parties give the impression that they are at loggerheads already the arbitrator will be able to anticipate extensive procedural wrangles and in those circumstances he would be sensible seriously to consider setting a limit on recoverable costs. A percentage of the amount in dispute might be appropriate. It may be that in any event he would come to the conclusion that it is inappropriate for a losing party to be required to reimburse the other side's costs that are greater than a certain percentage of the claim.

Figure 6.13 is an example of an order limiting recoverable costs.

FIGURE 6.13 – AN ORDER LIMITING RECOVERABLE COSTS

In the Matter of the Arbitration Act 1996

and

In the Matter of an Arbitration

between

[*Name of Claimant*]   CLAIMANT

and

[*Name of Respondent*]   RESPONDENT

ORDER No. 2
ORDER LIMITING RECOVERABLE COSTS

At the preliminary meeting held on [*date*] I informed the parties that in view of the amount in dispute in this arbitration I proposed to exercise my discretion to limit the recoverable costs of the arbitration as I may under section 65 of the Arbitration Act 1996.

The parties wished to make submissions to me on the need for me to limit these costs and on the level at which I should set that limit.

The parties have made those submissions and I make the following Orders:

1. The recoverable costs of this arbitration shall be set at [£...] excluding VAT. This excludes my own fees and expenses.
2. The costs of this Order shall be costs in the arbitration.

[*Signed*]
Arbitrator
[*Date*]

The authors consider that it is inappropriate to make this order as a percentage of a claim that has yet to be made as this runs the risk of the claimant inflating the claim unnecessarily. If the arbitrator considers that the limit of recoverable costs should relate in percentage terms to the amount claimed, such cynical inflation can be avoided by using as a basis the figure given by the claimant at the preliminary meeting as the amount of the claim.

6.6.5 Costs of interlocutory proceedings

Often during the interlocutory stage, various applications from the parties will have to be dealt with, sometimes by way of a short hearing. While in many cases the costs arising from such applications may be dealt with as part of the general costs of the arbitration, they sometimes have to be considered separately. For instance, a party may apply for leave to amend his pleadings and this may be resisted by the other party or a party may consider that further and better particulars of the other's pleadings are required and this also may be resisted.

In cases such as this, the costs arising should be separately considered because it may not be fair, in the end, for the party ordered to pay the general costs to pay those special costs as well. In the first example, it is customary for the costs of an application to amend pleadings to be paid by the applicant in any event, unless the other party's resistance is totally unreasonable. Similarly, in the second example, whoever 'wins' on the application should not have to pay any of the costs. In such cases, the interlocutory order resulting from the application should deal with the costs involved, e.g. by saying 'The Claimant shall pay the costs of and incidental to the application, the hearing and this order in any event' and the arbitrator must take account of this when making the final award of costs. It is customary for such costs not actually to be paid at the time of the application, unless the arbitrator specifically directs otherwise, which is unusual.

6.6.7 Sealed and other offers

The 'sealed offer' in arbitration is the equivalent of a payment into court. The purpose of the sealed offer is to limit the offeror's liability to pay the other side's costs. The rule generally is that the offer, which must be an open offer, and not one without prejudice, is placed in a sealed envelope and given to the arbitrator at the conclusion of the hearing with a request that it be opened only after he has decided upon his award on the substantive issues. If, on opening the envelope, he finds that the amount offered is less than he has awarded, he will disregard it. If it is the same or more than his award, then he will generally award costs following the date of the offer in favour of the offeror, since, had the offer been accepted, the arbitration need not have proceeded beyond that date.

The offer should offer to pay the costs up to the time of its receipt

and state whether or not it includes interest. It should not be conditional in any way, e.g., it should not require acceptance within a
specified time.

There are a number of objections to this practice, the principal one
being that the arbitrator is made aware that an offer has been made,
even though he does not know the amount. But the practice has
received judicial support, notably in *Tramoutana Armadora SA* v.
Atlantic Shipping Co SA [1978] 2 All ER 870; 1 Lloyd's Rep 391, where
the following principle was laid down:

> 'If the claimant in the end has achieved no more than he would
> have achieved by accepting the offer, the continuance of the
> arbitration after that date has been a waste of time and money.
> *Prima facie*, the claimant should recover his costs up to the date of
> the offer and should be ordered to pay the respondent's costs
> after that date.'

In the same case, an alternative was suggested, namely that the
arbitrator could invite the respondent, at the end of the hearing, to
give him a sealed envelope containing either a statement that a
sealed offer had not been made or the sealed offer itself. There are a
number of variations on the theme. For instance, the arbitrator may
be asked to reserve his award on costs until he has handed down his
partial award on the issues, when the parties can address him as to
costs and the existence of the offer may then be revealed. The offer
will then often be in the form which has become known as a 'Calderbank letter', that is a letter containing an offer and stated to be
'without prejudice save as to costs', this prevents the letter being
produced by the offeree in the hearing of the substantive issues, but
allows the offeror to produce it at the hearing on costs.

A Calderbank offer should not include the costs themselves
which must be considered separately. If it does state that it is to
cover costs as well, it does not fulfil the requirements of this process.

In a sealed offer or Calderbank letter, interest and costs should
not be included in the sum offered as this can create doubt as to
whether the offer has been equalled or exceeded in the event and
may render the offer ineffective. The offer should be expressed as
'£x in full and final settlement of all claims and counterclaims plus
interest and costs to the date of this offer'.

Chapter Seven
After the Award

7.1 Introduction

Once he has completed his last award the arbitrator is generally considered to have completed his task. He is *functus officio* – a Latin phrase meaning that he has discharged his duties and has no further function to perform. He cannot, therefore, rescind his award and hear the case afresh. His authority as arbitrator has ended.

7.2 Corrections and new evidence

There are however certain circumstances where the arbitrator can correct an award or make an additional award and these are dealt with in section 57 of the 1996 Act.

As with many of the provisions in the 1996 Act the parties are free to agree on the powers of the arbitrator in this respect. This means that the parties can agree anything between a situation where the arbitrator has no powers whatsoever to correct an award to the complete opposite where he could do anything he likes to his award after it has been made.

Neither of these situations is particularly attractive and section 57 as with most other sections of the 1996 Act contains provisions that will apply should the parties not reach an agreement in this respect and which are perhaps the best compromise for most situations.

The provisions included in section 57 in the absence of agreement otherwise are as follows:

- the arbitrator may correct an award so as to remove any clerical mistake or error arising from an accidental slip or omission or clarify or remove any ambiguity in the award, or
- he may make an additional award in respect of any claim (including a claim for interest or costs) which was presented to him but was not dealt with in the award.

These actions can be taken on the arbitrator's own initiative or on the application of a party but the other party, or both parties in the case of an arbitrator taking the initiative in this respect, must have a reasonable opportunity to make representations to the arbitrator. The time limits for carrying out these actions are very limited unless the parties agree to extend the period set in the 1996 Act.

One situation that may arise is where fresh evidence appears after the award has been made. There is nothing in the 1996 Act permitting a party finding new evidence after the award has been made to apply to the court for the award to be remitted back to the arbitrator for consideration of that evidence as was formerly possible under section 22 of the 1950 Act; see *Re Keighley Maxstead & Co and Durant & Co* [1891–1894] All ER 1012, a decision of the Court of Appeal. It would therefore appear that a party would have to apply to the arbitrator to re-open the arbitration to deal with the new evidence and this application could well be opposed on the basis that the arbitrator is no longer in office. At the time of writing we await with interest the first case where this is tested.

The party requiring that the new evidence be considered would have to apply to the court for the award to be remitted to the arbitrator when, even though *functus officio*, he would effectively be reinstated.

7.3   Enforcing the award

An award may be enforced by the successful party in two ways. First, and most conveniently, he may apply to the court for leave for the award to be enforced in the same manner as a judgment or order of the court to the same effect as provided by section 66(1) of the 1996 Act. Where such leave is given judgment may be entered in terms of the award and it is then enforceable in the same way as a judgment of the court.

The second method of enforcing an award is to sue on it, but this is not generally advisable and is seldom done in practice. The action is one in contract for breach of an implied term that any award made will be honoured. A breach of that implied term arising out of the failure to honour the award gives rise to an independent cause of action distinct from the original breach of contract which gave rise to, and was the subject matter of the arbitration. It has been held that the time limit laid down in section 7 of the Limitation Act 1980, namely that 'an action to enforce an award' must be brought within six years from the date on which 'the cause of action accrued' begins to run from breach of the implied term to perform the award and not from the date of the original cause of action: *Agromet Motoimport Ltd* v. *Maulden Engineering Co (Beds) Ltd* (1985) 2 All ER 436.

Section 66(3) covers the situation where it is shown that the arbitrator lacked substantive jurisdiction to make the award. In that event the award will not be enforced.

7.4 Challenges to the award

In addition to its powers of enforcement the court also deals with challenges to the award. These can be made on the following bases

- a lack of substantive jurisdiction on the part of the tribunal, (section 67 of the 1996 Act);
- a serious irregularity affecting the tribunal (section 68 of the 1996 Act).

A party can also appeal to the court on a question of law arising out of an award.

7.4.1 Lack of substantive jurisdiction

There are two instances where a challenge to the award on the basis of the arbitrator lacking substantive jurisdiction can be made to the court under section 67 of the 1996 Act. These are:

- where the tribunal has made an award in accordance with section 31(4)(a) in respect of its own substantive jurisdiction;
- where the tribunal is alleged not to have had substantive jurisdiction in respect of the whole or a part of the award on the merits.

The court has three options set out in section 67(3) when it deals with a challenge to an award on the basis of lack of substantive jurisdiction. These are:

- to confirm the award
- to vary the award, or
- to set the award aside in whole or in part.

It is noteworthy that there is no provision for the award to be remitted to the arbitrator in these circumstances and where the award is set aside it would appear that unless the parties are prepared to agree that the arbitrator's jurisdiction can be extended, the arbitration has to start again from square one.

There is no indication in the 1996 Act as to the position of the arbitrator himself in these circumstances. If he has made his final award he is, as indicated previously, *functus officio* and will thus have no further involvement in the proceedings. If however the award in question is a partial award, the arbitrator is still in post and if the parties do not agree to his jurisdiction being extended he may have no alternative but to resign. He would then have to agree with the parties the consequences of his resignation in accordance with the requirements of section 25 of the 1996 Act as regards his entitlement (if any) to fees or expenses or any liability incurred by him as a result of his resignation.

7.4.2 Serious irregularities

Section 68(2) of the 1996 Act sets out a list of serious irregularities which an arbitrator can be accused of committing. These are:

- failure to comply with section 33, (the obligations to act fairly and impartially as between the parties and to adopt procedures suitable to the circumstances of the case avoiding unnecessary delay or expense;
- the arbitrator exceeding his powers (other than by exceeding his substantive jurisdiction);
- failure by the arbitrator to conduct the proceedings in accordance with the procedure agreed between the parties;
- failure to deal with all the issues that were put to arbitration;
- any arbitral or other institution or person vested by the parties with powers relating to the proceedings or the award exceeding its powers;
- uncertainty or ambiguity as to the effect of the award;
- the award being obtained by fraud or the award or the way in which it was procured being contrary to public policy;
- failure to comply with the requirements as to the form of the award;
- any admitted irregularity in the conduct of the proceedings or in the award.

In the same way as in the case of an application under section 67 the court has three options as to the action it may take when faced with an application regarding an allegation of a serious irregularity perpetrated by the arbitrator. These options do however only include one that is the same as those applying in the case of a challenge to jurisdiction under section 67. The court's options in this instance are:

- to remit the award to the arbitrator in whole or in part for reconsideration,
- to set the award aside in whole or in part, or
- to declare the award to be of no effect, in whole or in part.

Section 68(3) continues as follows:

'The court shall not exercise its power to set aside or declare an award to be of no effect, in whole or in part, unless it is satisfied that it would be inappropriate to remit the matters in question to the tribunal [i.e. the arbitrator] for reconsideration.'

Where the award is remitted to the arbitrator, the question may arise in the arbitrator's mind as to his entitlement to fees for the reconsideration. This is obviously very much a matter for the individual circumstances to dictate but as a general rule, if the arbitrator is of the view that the remission has resulted from his own omission, he should be prepared to make the necessary revisions to his award free of charge to the parties and by so doing avoid a possible challenge in the courts of any fee he might seek to charge.

Where an award is set aside or declared to be of no effect, as far as the arbitrator is concerned the situation is the same as that described in the case of the setting aside of an award for lack of substantive jurisdiction as considered above.

7.5 *Appeals on a point of law*

Section 69 of the 1996 Act allows a party, unless agreed otherwise, to appeal to the court on a question of law arising out of an award. This is the section, mentioned in 6.2.6, that is inoperative should the parties agree to dispense with reasons in the award.

All parties must agree to such an appeal being made or the leave of the court must be obtained.

When an appeal of this kind is made there are four actions that the court may take having heard the parties' submissions. These are:

- to confirm the award
- to vary the award
- to remit the award to the arbitrator in whole or in part, for reconsideration in the light of the court's determination, or
- set aside the award in whole or in part.

The position of the parties and the arbitrator after the court has made its decision as to the action it should take in the event of an appeal on a point of law is the same as has been discussed above.

7.6 The power of the court to remove the arbitrator

The 1996 Act specifically avoids the use of the word misconduct when referring to actions by the arbitrator which do not find favour with the parties. Instead the words 'serious irregularity' are used. In earlier Arbitration Acts there was a close relationship between the setting aside of an arbitrator's award and the removal of the arbitrator for misconduct.

The 1996 Act avoids any mention of the removal of the arbitrator in section 68 which deals with challenging the award for serious irregularities.

The removal of the arbitrator is dealt with in section 24 of the 1996 Act. An action to remove the arbitrator can arise at any time during the course of the arbitration and the grounds for such an action are set out in detail in section 24. They are:

- '(a) that circumstances exist that give rise to justifiable doubts as to his impartiality;'

If the arbitrator follows the requirements of section 33(a) and acts fairly and impartially as between the parties, giving each party a reasonable opportunity of putting his case and dealing with that of his opponent, he should have little to fear in this regard. He should of course also ensure that there can be no possibility that a reasonable man would consider any of his actions biased in favour of one party only.

- '(b) that he does not possess the qualifications required by the arbitration agreement;'

It is of course essential that an arbitrator before accepting appointment ensures that he does fit any requirements set out in the arbitration agreement.

185

- '(c) that he is physically or mentally incapable of conducting the proceedings or there are justifiable doubts as to his capacity to do so;'
- '(d) that he has refused or failed –
    (i) properly to conduct the proceedings, or
    (ii) to use all reasonable despatch in conducting the proceedings or making an award,
  and that substantial injustice has been or will be caused to the applicant.

This last is the sting in the tail! An accusation that an arbitrator has refused or failed properly to conduct the proceedings can be raised by a party who can find no other reason to contest an award made against him.

The DAC Report is however quite clear that the ground set out in section 24(1)(d)

'only exists to cover what we hope will be the very rare case where an arbitrator so conducts the proceedings that it can be fairly said that instead of carrying through the object of arbitration as stated in the Act, he is in effect frustrating that object.'

As there is no automatic connection between the provisions of section 68 and section 24, appeals to the court under the 1996 Act in respect of allegations of failures on the part of arbitrators will be limited to those circumstances where it is obvious that the arbitrator is by his actions frustrating the object set out in section 1(a) of the 1996 Act, that is,

'to obtain the fair resolution of disputes by an impartial tribunal without unnecessary delay or expense'.

If a party seeks to disrupt the proceedings by making unmeritorious applications to the court for the removal of an arbitrator it is to be hoped that the court will give them short shrift and thereby fulfil the desires of the DAC.

In this respect it is worth mentioning that the DAC Report was cited with authority and relied upon when the Arbitration Bill was being debated in the House of Lords. (*See the Hansard Reports of 28th February 1996, CWH 1 – CWH 30 which relate to the committee stage, and of 18th March 1996, 1075–1087 relating to the debate in the House itself.*)

During the debate in the committee stage the case of *Pepper (Inspector of Taxes)* v. *Hart* [1993] AC 293, itself a decision of the

House of Lords, was quoted. In this case it was held that where there is ambiguity as to the meaning of a section within a statute, it is perfectly proper to refer to the Hansard Reports on that statute in order to determine the true intention of the legislator. Whether the courts themselves will treat the DAC Report as being elevated to the status of authority on the Act is however, at the time of writing, not clear.

The arbitrators' actions that are identified in the 'cautionary tales' set out at the end of Chapter 3 are worthy of a reminder even though they applied to arbitrations run under the 1950 and 1979 Acts. Actions such as the ones perpetrated in those cases would still be likely to fall foul of the provisions of section 24 of the 1996 Act.

7.7 Finally

The time will ultimately arrive when the arbitrator is satisfied that he has completed his duties. As set out in Chapter 4 he should keep his notes and correspondence files for six years or even longer. He will also have accumulated the hearing bundles, witness statements, experts' reports and possibly a large number of sundry documents. A useful procedure is to request the parties to collect these documents within a given time, say 28 days, or they will be destroyed. As parties seldom avail themselves of this offer an indispensible tool for the arbitrator is a large garden incinerator.

Appendix 1
The Arbitration Act 1996

ARRANGEMENT OF SECTIONS

Section

Construction Arbitrations

An Act to restate and improve the law relating to arbitration pursuant to an arbitration agreement; to make other provision relating to arbitration and arbitration awards; and for connected purposes. [17th June 1996]

Part I

Arbitration pursuant to an arbitration agreement

Introductory

General principles.

1. The provisions of this Part are founded on the following principles, and shall be construed accordingly —

    (a) the object of arbitration is to obtain the fair resolution of disputes by an impartial tribunal without unnecessary delay or expense;

    (b) the parties should be free to agree how their disputes are resolved, subject only to such safeguards as are necessary in the public interest;

    (c) in matters governed by this Part the court should not intervene except as provided by this Part.

Scope of application of provisions.

2. — (1) The provisions of this Part apply where the seat of the arbitration is in England and Wales or Northern Ireland.

(2) The following sections apply even if the seat of the arbitration is outside England and Wales or Northern Ireland or no seat has been designated or determined —

    (a) sections 9 to 11 (stay of legal proceedings, &c.), and

    (b) section 66 (enforcement of arbitral awards).

(3) The powers conferred by the following sections apply even if the seat of the arbitration is outside England and Wales or Northern Ireland or no seat has been designated or determined —

    (a) section 43 (securing the attendance of witnesses), and

    (b) section 44 (court powers exercisable in support of arbitral proceedings);

but the court may refuse to exercise any such power if, in the opinion of the court, the fact that the seat of the arbitration is outside England and Wales or Northern Ireland, or that when designated or determined the seat is likely to be outside England and Wales or Northern Ireland, makes it inappropriate to do so.

(4) The court may exercise a power conferred by any provision of this Part not mentioned in subsection (2) or (3) for the purpose of supporting the arbitral process where —

    (a) no seat of the arbitration has been designated or determined, and

    (b) by reason of a connection with England and Wales or Northern Ireland the court is satisfied that it is appropriate to do so.

(5) Section 7 (separability of arbitration agreement) and section 8 (death of a party) apply where the law applicable to the arbitration agreement is the law of England and Wales or Northern Ireland even if the seat of the

arbitration is outside England and Wales or Northern Ireland or has not been designated or determined.

3. In this Part "the seat of the arbitration" means the juridical seat of the arbitration designated—

    (a) by the parties to the arbitration agreement, or

    (b) by any arbitral or other institution or person vested by the parties with powers in that regard, or

    (c) by the arbitral tribunal if so authorised by the parties,

or determined, in the absence of any such designation, having regard to the parties' agreement and all the relevant circumstances.

The seat of the arbitration.

4.—(1) The mandatory provisions of this Part are listed in Schedule 1 and have effect notwithstanding any agreement to the contrary.

(2) The other provisions of this Part (the "non-mandatory provisions") allow the parties to make their own arrangements by agreement but provide rules which apply in the absence of such agreement.

(3) The parties may make such arrangements by agreeing to the application of institutional rules or providing any other means by which a matter may be decided.

(4) It is immaterial whether or not the law applicable to the parties' agreement is the law of England and Wales or, as the case may be, Northern Ireland.

(5) The choice of a law other than the law of England and Wales or Northern Ireland as the applicable law in respect of a matter provided for by a non-mandatory provision of this Part is equivalent to an agreement making provision about that matter.

For this purpose an applicable law determined in accordance with the parties' agreement, or which is objectively determined in the absence of any express or implied choice, shall be treated as chosen by the parties.

Mandatory and non-mandatory provisions.

5.—(1) The provisions of this Part apply only where the arbitration agreement is in writing, and any other agreement between the parties as to any matter is effective for the purposes of this Part only if in writing.

The expressions "agreement", "agree" and "agreed" shall be construed accordingly.

(2) There is an agreement in writing—

    (a) if the agreement is made in writing (whether or not it is signed by the parties),

    (b) if the agreement is made by exchange of communications in writing, or

    (c) if the agreement is evidenced in writing.

Agreements to be in writing.

(3) Where parties agree otherwise than in writing by reference to terms which are in writing, they make an agreement in writing.

(4) An agreement is evidenced in writing if an agreement made otherwise than in writing is recorded by one of the parties, or by a third party, with the authority of the parties to the agreement.

(5) An exchange of written submissions in arbitral or legal proceedings in which the existence of an agreement otherwise than in writing is alleged by one party against another party and not denied by the other party in his response constitutes as between those parties an agreement in writing to the effect alleged.

(6) References in this Part to anything being written or in writing include its being recorded by any means.

The arbitration agreement

Definition of arbitration agreement.

6. — (1) In this Part an "arbitration agreement" means an agreement to submit to arbitration present or future disputes (whether they are contractual or not).

(2) The reference in an agreement to a written form of arbitration clause or to a document containing an arbitration clause constitutes an arbitration agreement if the reference is such as to make that clause part of the agreement.

Separability of arbitration agreement.

7. Unless otherwise agreed by the parties, an arbitration agreement which forms or was intended to form part of another agreement (whether or not in writing) shall not be regarded as invalid, non-existent or ineffective because that other agreement is invalid, or did not come into existence or has become ineffective, and it shall for that purpose be treated as a distinct agreement.

Whether agreement discharged by death of a party.

8. — (1) Unless otherwise agreed by the parties, an arbitration agreement is not discharged by the death of a party and may be enforced by or against the personal representatives of that party.

(2) Subsection (1) does not affect the operation of any enactment or rule of law by virtue of which a substantive right or obligation is extinguished by death.

Stay of legal proceedings

Stay of legal proceedings.

9. — (1) A party to an arbitration agreement against whom legal proceedings are brought (whether by way of claim or counterclaim) in respect of a matter which under the agreement is to be referred to arbitration may (upon notice to the other parties to the proceedings) apply to the court in which the proceedings have been brought to stay the proceedings so far as they concern that matter.

(2) An application may be made notwithstanding that the matter is to be referred to arbitration only after the exhaustion of other dispute resolution procedures.

(3) An application may not be made by a person before taking the appropriate procedural step (if any) to acknowledge the legal proceedings against him or after he has taken any step in those proceedings to answer the substantive claim.

(4) On an application under this section the court shall grant a stay unless satisfied that the arbitration agreement is null and void, inoperative, or incapable of being performed.

(5) If the court refuses to stay the legal proceedings, any provision that an award is a condition precedent to the bringing of legal proceedings in respect of any matter is of no effect in relation to those proceedings.

10.—(1) Where in legal proceedings relief by way of interpleader is granted and any issue between the claimants is one in respect of which there is an arbitration agreement between them, the court granting the relief shall direct that the issue be determined in accordance with the agreement unless the circumstances are such that proceedings brought by a claimant in respect of the matter would not be stayed. *Reference of interpleader issue to arbitration.*

(2) Where subsection (1) applies but the court does not direct that the issue be determined in accordance with the arbitration agreement, any provision that an award is a condition precedent to the bringing of legal proceedings in respect of any matter shall not affect the determination of that issue by the court.

11.—(1) Where Admiralty proceedings are stayed on the ground that the dispute in question should be submitted to arbitration, the court granting the stay may, if in those proceedings property has been arrested or bail or other security has been given to prevent or obtain release from arrest— *Retention of security where Admiralty proceedings stayed.*
  (a) order that the property arrested be retained as security for the satisfaction of any award given in the arbitration in respect of that dispute, or
  (b) order that the stay of those proceedings be conditional on the provision of equivalent security for the satisfaction of any such award.

(2) Subject to any provision made by rules of court and to any necessary modifications, the same law and practice shall apply in relation to property retained in pursuance of an order as would apply if it were held for the purposes of proceedings in the court making the order.

Commencement of arbitral proceedings

12.—(1) Where an arbitration agreement to refer future disputes to arbitration provides that a claim shall be barred, or the claimant's right *Power of court to extend time for beginning arbitral proceedings, &c.*

195

extinguished, unless the claimant takes within a time fixed by the agreement some step —

    (a) to begin arbitral proceedings, or

    (b) to begin other dispute resolution procedures which must be exhausted before arbitral proceedings can be begun,

the court may by order extend the time for taking that step.

(2) Any party to the arbitration agreement may apply for such an order (upon notice to the other parties), but only after a claim has arisen and after exhausting any available arbitral process for obtaining an extension of time.

(3) The court shall make an order only if satisfied —

    (a) that the circumstances are such as were outside the reasonable contemplation of the parties when they agreed the provision in question, and that it would be just to extend the time, or

    (b) that the conduct of one party makes it unjust to hold the other party to the strict terms of the provision in question.

(4) The court may extend the time for such period and on such terms as it thinks fit, and may do so whether or not the time previously fixed (by agreement or by a previous order) has expired.

(5) An order under this section does not affect the operation of the Limitation Acts (see section 13).

(6) The leave of the court is required for any appeal from a decision of the court under this section.

Application of Limitation Acts.
    13. — (1) The Limitation Acts apply to arbitral proceedings as they apply to legal proceedings.

(2) The court may order that in computing the time prescribed by the Limitation Acts for the commencement of proceedings (including arbitral proceedings) in respect of a dispute which was the subject matter —

    (a) of an award which the court orders to be set aside or declares to be of no effect, or

    (b) of the affected part of an award which the court orders to be set aside in part, or declares to be in part of no effect,

the period between the commencement of the arbitration and the date of the order referred to in paragraph (a) or (b) shall be excluded.

(3) In determining for the purposes of the Limitation Acts when a cause of action accrued, any provision that an award is a condition precedent to the bringing of legal proceedings in respect of a matter to which an arbitration agreement applies shall be disregarded.

(4) In this Part "the Limitation Acts" means —

    (a) in England and Wales, the Limitation Act 1980, the Foreign Limitation Periods Act 1984 and any other enactment (whenever passed) relating to the limitation of actions;

(b) in Northern Ireland, the Limitation (Northern Ireland) Order 1989, the Foreign Limitation Periods (Northern Ireland) Order 1985 and any other enactment (whenever passed) relating to the limitation of actions.

14.—(1) The parties are free to agree when arbitral proceedings are to be regarded as commenced for the purposes of this Part and for the purposes of the Limitation Acts.

Commencement
of arbitral
proceedings.

(2) If there is no such agreement the following provisions apply.

(3) Where the arbitrator is named or designated in the arbitration agreement, arbitral proceedings are commenced in respect of a matter when one party serves on the other party or parties a notice in writing requiring him or them to submit that matter to the person so named or designated.

(4) Where the arbitrator or arbitrators are to be appointed by the parties, arbitral proceedings are commenced in respect of a matter when one party serves on the other party or parties notice in writing requiring him or them to appoint an arbitrator or to agree to the appointment of an arbitrator in respect of that matter.

(5) Where the arbitrator or arbitrators are to be appointed by a person other than a party to the proceedings, arbitral proceedings are commenced in respect of a matter when one party gives notice in writing to that person requesting him to make the appointment in respect of that matter.

The arbitral tribunal

15.—(1) The parties are free to agree on the number of arbitrators to form the tribunal and whether there is to be a chairman or umpire.

The arbitral
tribunal.

(2) Unless otherwise agreed by the parties, an agreement that the number of arbitrators shall be two or any other even number shall be understood as requiring the appointment of an additional arbitrator as chairman of the tribunal.

(3) If there is no agreement as to the number of arbitrators, the tribunal shall consist of a sole arbitrator.

16.—(1) The parties are free to agree on the procedure for appointing the arbitrator or arbitrators, including the procedure for appointing any chairman or umpire.

Procedure for
appointment of
arbitrators.

(2) If or to the extent that there is no such agreement, the following provisions apply.

(3) If the tribunal is to consist of a sole arbitrator, the parties shall jointly appoint the arbitrator not later than 28 days after service of a request in writing by either party to do so.

(4) If the tribunal is to consist of two arbitrators, each party shall appoint one arbitrator not later than 14 days after service of a request in writing by either party to do so.

(5) If the tribunal is to consist of three arbitrators —
- (a) each party shall appoint one arbitrator not later than 14 days after service of a request in writing by either party to do so, and
- (b) the two so appointed shall forthwith appoint a third arbitrator as the chairman of the tribunal.

(6) If the tribunal is to consist of two arbitrators and an umpire —
- (a) each party shall appoint one arbitrator not later than 14 days after service of a request in writing by either party to do so, and
- (b) the two so appointed may appoint an umpire at any time after they themselves are appointed and shall do so before any substantive hearing or forthwith if they cannot agree on a matter relating to the arbitration.

(7) In any other case (in particular, if there are more than two parties) section 18 applies as in the case of a failure of the agreed appointment procedure.

Power in case of default to appoint sole arbitrator.

17. — (1) Unless the parties otherwise agree, where each of two parties to an arbitration agreement is to appoint an arbitrator and one party ("the party in default") refuses to do so, or fails to do so within the time specified, the other party, having duly appointed his arbitrator, may give notice in writing to the party in default that he proposes to appoint his arbitrator to act as sole arbitrator.

(2) If the party in default does not within 7 clear days of that notice being given —
- (a) make the required appointment, and
- (b) notify the other party that he has done so,

the other party may appoint his arbitrator as sole arbitrator whose award shall be binding on both parties as if he had been so appointed by agreement.

(3) Where a sole arbitrator has been appointed under subsection (2), the party in default may (upon notice to the appointing party) apply to the court which may set aside the appointment.

(4) The leave of the court is required for any appeal from a decision of the court under this section.

Failure of appointment procedure.

18. — (1) The parties are free to agree what is to happen in the event of a failure of the procedure for the appointment of the arbitral tribunal.

There is no failure if an appointment is duly made under section 17 (power in case of default to appoint sole arbitrator), unless that appointment is set aside.

(2) If or to the extent that there is no such agreement any party to the arbitration agreement may (upon notice to the other parties) apply to the court to exercise its powers under this section.

(3) Those powers are—
    (a) to give directions as to the making of any necessary appointments;
    (b) to direct that the tribunal shall be constituted by such appointments (or any one or more of them) as have been made;
    (c) to revoke any appointments already made;
    (d) to make any necessary appointments itself.

(4) An appointment made by the court under this section has effect as if made with the agreement of the parties.

(5) The leave of the court is required for any appeal from a decision of the court under this section.

19. In deciding whether to exercise, and in considering how to exercise, any of its powers under section 16 (procedure for appointment of arbitrators) or section 18 (failure of appointment procedure), the court shall have due regard to any agreement of the parties as to the qualifications required of the arbitrators.

Court to have regard to agreed qualifications.

20.—(1) Where the parties have agreed that there is to be a chairman, they are free to agree what the functions of the chairman are to be in relation to the making of decisions, orders and awards.

Chairman.

(2) If or to the extent that there is no such agreement, the following provisions apply.

(3) Decisions, orders and awards shall be made by all or a majority of the arbitrators (including the chairman).

(4) The view of the chairman shall prevail in relation to a decision, order or award in respect of which there is neither unanimity nor a majority under subsection (3).

21.—(1) Where the parties have agreed that there is to be an umpire, they are free to agree what the functions of the umpire are to be, and in particular—
    (a) whether he is to attend the proceedings, and
    (b) when he is to replace the other arbitrators as the tribunal with power to make decisions, orders and awards.

Umpire.

(2) If or to the extent that there is no such agreement, the following provisions apply.

(3) The umpire shall attend the proceedings and be supplied with the same documents and other materials as are supplied to the other arbitrators.

(4) Decisions, orders and awards shall by made by the other arbitrators unless and until they cannot agree on a matter relating to the arbitration.

In that event they shall forthwith give notice in writing to the parties and the umpire, whereupon the umpire shall replace them as the tribunal with power to make decisions, orders and awards as if he were sole arbitrator.

(5) If the arbitrators cannot agree but fail to give notice of that fact, or if any of them fails to join in the giving of notice, any party to the arbitral proceedings may (upon notice to the other parties and to the tribunal) apply to the court which may order that the umpire shall replace the other arbitrators as the tribunal with power to make decisions, orders and awards as if he were sole arbitrator.

(6) The leave of the court is required for any appeal from a decision of the court under this section.

Decision-making where no chairman or umpire.

22. — (1) Where the parties agree that there shall be two or more arbitrators with no chairman or umpire, the parties are free to agree how the tribunal is to make decisions, orders and awards.

(2) If there is no such agreement, decisions, orders and awards shall be made by all or a majority of the arbitrators.

Revocation of arbitrator's authority.

23. — (1) The parties are free to agree in what circumstances the authority of an arbitrator may be revoked.

(2) If or to the extent that there is no such agreement the following provisions apply.

(3) The authority of an arbitrator may not be revoked except —
    (a) by the parties acting jointly, or
    (b) by an arbitral or other institution or person vested by the parties with powers in that regard.

(4) Revocation of the authority of an arbitrator by the parties acting jointly must be agreed in writing unless the parties also agree (whether or not in writing) to terminate the arbitration agreement.

(5) Nothing in this section affects the power of the court —
    (a) to revoke an appointment under section 18 (powers exercisable in case of failure of appointment procedure), or
    (b) to remove an arbitrator on the grounds specified in section 24.

Power of court to remove arbitrator.

24. — (1) A party to arbitral proceedings may (upon notice to the other parties, to the arbitrator concerned and to any other arbitrator) apply to the court to remove an arbitrator on any of the following grounds —
    (a) that circumstances exist that give rise to justifiable doubts as to his impartiality;
    (b) that he does not possess the qualifications required by the arbitration agreement;

(c) that he is physically or mentally incapable of conducting the proceedings or there are justifiable doubts as to his capacity to do so;

(d) that he has refused or failed—

    (i) properly to conduct the proceedings, or

    (ii) to use all reasonable despatch in conducting the proceedings or making an award,

and that substantial injustice has been or will be caused to the applicant.

(2) If there is an arbitral or other institution or person vested by the parties with power to remove an arbitrator, the court shall not exercise its power of removal unless satisfied that the applicant has first exhausted any available recourse to that institution or person.

(3) The arbitral tribunal may continue the arbitral proceedings and make an award while an application to the court under this section is pending.

(4) Where the court removes an arbitrator, it may make such order as it thinks fit with respect to his entitlement (if any) to fees or expenses, or the repayment of any fees or expenses already paid.

(5) The arbitrator concerned is entitled to appear and be heard by the court before it makes any order under this section.

(6) The leave of the court is required for any appeal from a decision of the court under this section.

25.—(1) The parties are free to agree with an arbitrator as to the consequences of his resignation as regards— **Resignation of arbitrator.**

(a) his entitlement (if any) to fees or expenses, and

(b) any liability thereby incurred by him.

(2) If or to the extent that there is no such agreement the following provisions apply.

(3) An arbitrator who resigns his appointment may (upon notice to the parties) apply to the court—

(a) to grant him relief from any liability thereby incurred by him, and

(b) to make such order as it thinks fit with respect to his entitlement (if any) to fees or expenses or the repayment of any fees or expenses already paid.

(4) If the court is satisfied that in all the circumstances it was reasonable for the arbitrator to resign, it may grant such relief as is mentioned in subsection (3)(a) on such terms as it thinks fit.

(5) The leave of the court is required for any appeal from a decision of the court under this section.

Death of arbitrator or person appointing him.

26. – (1) The authority of an arbitrator is personal and ceases on his death.

(2) Unless otherwise agreed by the parties, the death of the person by whom an arbitrator was appointed does not revoke the arbitrator's authority.

Filling of vacancy, &c.

27. – (1) Where an arbitrator ceases to hold office, the parties are free to agree –

(a) whether and if so how the vacancy is to be filled,

(b) whether and if so to what extent the previous proceedings should stand, and

(c) what effect (if any) his ceasing to hold office has on any appointment made by him (alone or jointly).

(2) If or to the extent that there is no such agreement, the following provisions apply.

(3) The provisions of sections 16 (procedure for appointment of arbitrators) and 18 (failure of appointment procedure) apply in relation to the filling of the vacancy as in relation to an original appointment.

(4) The tribunal (when reconstituted) shall determine whether and if so to what extent the previous proceedings should stand.

This does not affect any right of a party to challenge those proceedings on any ground which had arisen before the arbitrator ceased to hold office.

(5) His ceasing to hold office does not affect any appointment by him (alone or jointly) of another arbitrator, in particular any appointment of a chairman or umpire.

Joint and several liability of parties to arbitrators for fees and expenses.

28. – (1) The parties are jointly and severally liable to pay to the arbitrators such reasonable fees and expenses (if any) as are appropriate in the circumstances.

(2) Any party may apply to the court (upon notice to the other parties and to the arbitrators) which may order that the amount of the arbitrators' fees and expenses shall be considered and adjusted by such means and upon such terms as it may direct.

(3) If the application is made after any amount has been paid to the arbitrators by way of fees or expenses, the court may order the repayment of such amount (if any) as is shown to be excessive, but shall not do so unless it is shown that it is reasonable in the circumstances to order repayment.

(4) The above provisions have effect subject to any order of the court under section 24(4) or 25(3)(b) (order as to entitlement to fees or expenses in case of removal or resignation of arbitrator).

(5) Nothing in this section affects any liability of a party to any other

party to pay all or any of the costs of the arbitration (see sections 59 to 65) or any contractual right of an arbitrator to payment of his fees and expenses.

(6) In this section references to arbitrators include an arbitrator who has ceased to act and an umpire who has not replaced the other arbitrators.

29.—(1) An arbitrator is not liable for anything done or omitted in the discharge or purported discharge of his functions as arbitrator unless the act or omission is shown to have been in bad faith.

Immunity of arbitrator.

(2) Subsection (1) applies to an employee or agent of an arbitrator as it applies to the arbitrator himself.

(3) This section does not affect any liability incurred by an arbitrator by reason of his resigning (but see section 25).

Jurisdiction of the arbitral tribunal

30.—(1) Unless otherwise agreed by the parties, the arbitral tribunal may rule on its own substantive jurisdiction, that is, as to—

Competence of tribunal to rule on its own jurisdiction.

(a) whether there is a valid arbitration agreement,
(b) whether the tribunal is properly constituted, and
(c) what matters have been submitted to arbitration in accordance with the arbitration agreement.

(2) Any such ruling may be challenged by any available arbitral process of appeal or review or in accordance with the provisions of this Part.

31.—(1) An objection that the arbitral tribunal lacks substantive jurisdiction at the outset of the proceedings must be raised by a party not later than the time he takes the first step in the proceedings to contest the merits of any matter in relation to which he challenges the tribunal's jurisdiction.

Objection to substantive jurisdiction of tribunal.

A party is not precluded from raising such an objection by the fact that he has appointed or participated in the appointment of an arbitrator.

(2) Any objection during the course of the arbitral proceedings that the arbitral tribunal is exceeding its substantive jurisdiction must be made as soon as possible after the matter alleged to be beyond its jurisdiction is raised.

(3) The arbitral tribunal may admit an objection later than the time specified in subsection (1) or (2) if it considers the delay justified.

(4) Where an objection is duly taken to the tribunal's substantive jurisdiction and the tribunal has power to rule on its own jurisdiction, it may—

(a) rule on the matter in an award as to jurisdiction, or
(b) deal with the objection in its award on the merits.

If the parties agree which of these courses the tribunal should take, the tribunal shall proceed accordingly.

(5) The tribunal may in any case, and shall if the parties so agree, stay proceedings whilst an application is made to the court under section 32 (determination of preliminary point of jurisdiction).

Determination of preliminary point of jurisdiction.

32.—(1) The court may, on the application of a party to arbitral proceedings (upon notice to the other parties), determine any question as to the substantive jurisdiction of the tribunal.
A party may lose the right to object (see section 73).

(2) An application under this section shall not be considered unless—
- (a) it is made with the agreement in writing of all the other parties to the proceedings, or
- (b) it is made with the permission of the tribunal and the court is satisfied—
  - (i) that the determination of the question is likely to produce substantial savings in costs,
  - (ii) that the application was made without delay, and
  - (iii) that there is good reason why the matter should be decided by the court.

(3) An application under this section, unless made with the agreement of all the other parties to the proceedings, shall state the grounds on which it is said that the matter should be decided by the court.

(4) Unless otherwise agreed by the parties, the arbitral tribunal may continue the arbitral proceedings and make an award while an application to the court under this section is pending.

(5) Unless the court gives leave, no appeal lies from a decision of the court whether the conditions specified in subsection (2) are met.

(6) The decision of the court on the question of jurisdiction shall be treated as a judgment of the court for the purposes of an appeal.
But no appeal lies without the leave of the court which shall not be given unless the court considers that the question involves a point of law which is one of general importance or is one which for some other special reason should be considered by the Court of Appeal.

The arbitral proceedings

General duty of the tribunal.

33.—(1) The tribunal shall—
- (a) act fairly and impartially as between the parties, giving each party a reasonable opportunity of putting his case and dealing with that of his opponent, and
- (b) adopt procedures suitable to the circumstances of the particular case, avoiding unnecessary delay or expense, so as to provide a fair means for the resolution of the matters falling to be determined.

(2) The tribunal shall comply with that general duty in conducting the arbitral proceedings, in its decisions on matters of procedure and evidence and in the exercise of all other powers conferred on it.

34. — (1) It shall be for the tribunal to decide all procedural and evidential matters, subject to the right of the parties to agree any matter.

Procedural and evidential matters.

(2) Procedural and evidential matters include —

- (a) when and where any part of the proceedings is to be held;
- (b) the language or languages to be used in the proceedings and whether translations of any relevant documents are to be supplied;
- (c) whether any and if so what form of written statements of claim and defence are to be used, when these should be supplied and the extent to which such statements can be later amended;
- (d) whether any and if so which documents or classes of documents should be disclosed between and produced by the parties and at what stage;
- (e) whether any and if so what questions should be put to and answered by the respective parties and when and in what form this should be done;
- (f) whether to apply strict rules of evidence (or any other rules) as to the admissibility, relevance or weight of any material (oral, written or other) sought to be tendered on any matters of fact or opinion, and the time, manner and form in which such material should be exchanged and presented;
- (g) whether and to what extent the tribunal should itself take the initiative in ascertaining the facts and the law;
- (h) whether and to what extent there should be oral or written evidence or submissions.

(3) The tribunal may fix the time within which any directions given by it are to be complied with, and may if it thinks fit extend the time so fixed (whether or not it has expired).

35. — (1) The parties are free to agree —

- (a) that the arbitral proceedings shall be consolidated with other arbitral proceedings, or
- (b) that concurrent hearings shall be held, on such terms as may be agreed.

Consolidation of proceedings and concurrent hearings.

(2) Unless the parties agree to confer such power on the tribunal, the tribunal has no power to order consolidation of proceedings or concurrent hearings.

36. Unless otherwise agreed by the parties, a party to arbitral proceedings may be represented in the proceedings by a lawyer or other person chosen by him.

Legal or other representation.

Power to appoint experts, legal advisers or assessors.

37.—(1) Unless otherwise agreed by the parties—
    (a) the tribunal may—
        (i) appoint experts or legal advisers to report to it and the parties, or
        (ii) appoint assessors to assist it on technical matters,
        and may allow any such expert, legal adviser or assessor to attend the proceedings; and
    (b) the parties shall be given a reasonable opportunity to comment on any information, opinion or advice offered by any such person.

(2) The fees and expenses of an expert, legal adviser or assessor appointed by the tribunal for which the arbitrators are liable are expenses or the arbitrators for the purposes of this Part.

General powers exercisable by the tribunal.

38.—(1) The parties are free to agree on the powers exercisable by the arbitral tribunal for the purposes of and in relation to the proceedings.

(2) Unless otherwise agreed by the parties the tribunal has the following powers.

(3) The tribunal may order a claimant to provide security for the costs of the arbitration.
    This power shall not be exercised on the ground that the claimant is—
    (a) an individual ordinarily resident outside the United Kingdom, or
    (b) a corporation or association incorporated or formed under the law of a country outside the United Kingdom, or whose central management and control is exercised outside the United Kingdom.

(4) The tribunal may give directions in relation to any property which is the subject of the proceedings or as to which any question arises in the proceedings, and which is owned by or is in the possession of a party to the proceedings—
    (a) for the inspection, photographing, preservation, custody or detention of the property by the tribunal, an expert or a party, or
    (b) ordering that samples be taken from, or any observation be made of or experiment conducted upon, the property.

(5) The tribunal may direct that a party or witness shall be examined on oath or affirmation, and may for that purpose administer any necessary oath or take any necessary affirmation.

(6) The tribunal may give directions to a party for the preservation for the purposes of the proceedings of any evidence in his custody or control.

Power to make provisional awards.

39.—(1) The parties are free to agree that the tribunal shall have power to order on a provisional basis any relief which it would have power to grant in a final award.

(2) This includes, for instance, making—

(a) a provisional order for the payment of money or the disposition of property as between the parties, or

(b) an order to make an interim payment on account of the costs of the arbitration.

(3) Any such order shall be subject to the tribunal's final adjudication; and the tribunal's final award, on the merits or as to costs, shall take account of any such order.

(4) Unless the parties agree to confer such power on the tribunal, the tribunal has no such power.

This does not affect its powers under section 47 (awards on different issues, &c.).

40. — (1) The parties shall do all things necessary for the proper and expeditious conduct of the arbitral proceedings.

General duty of parties.

(2) This includes —
   (a) complying without delay with any determination of the tribunal as to procedural or evidential matters, or with any order or directions of the tribunal, and
   (b) where appropriate, taking without delay any necessary steps to obtain a decision of the court on a preliminary question of jurisdiction or law (see sections 32 and 45).

41. — (1) The parties are free to agree on the powers of the tribunal in case of a party's failure to do something necessary for the proper and expeditious conduct of the arbitration.

Powers of tribunal in case of party's default.

(2) Unless otherwise agreed by the parties, the following provisions apply.

(3) If the tribunal is satisfied that there has been inordinate and inexcusable delay on the part of the claimant in pursuing his claim and that the delay —
   (a) gives rise, or is likely to give rise, to a substantial risk that it is not possible to have a fair resolution of the issues in that claim, or
   (b) has caused, or is likely to cause, serious prejudice to the respondent,
the tribunal may make an award dismissing the claim.

(4) If without showing sufficient cause a party —
   (a) fails to attend or be represented at an oral hearing of which due notice was given, or
   (b) where matters are to be dealt with in writing, fails after due notice to submit written evidence or make written submissions,
the tribunal may continue the proceedings in the absence of that party or, as the case may be, without any written evidence or submissions on his behalf, and may make an award on the basis of the evidence before it.

(5) If without showing sufficient cause a party fails to comply with any order or directions of the tribunal, the tribunal may make a peremptory order to the same effect, prescribing such time for compliance with it as the tribunal considers appropriate.

(6) If a claimant fails to comply with a peremptory order of the tribunal to provide security for costs, the tribunal may make an award dismissing his claim.

(7) If a party fails to comply with any other kind of peremptory order, then, without prejudice to section 42 (enforcement by court of tribunal's peremptory orders), the tribunal may do any of the following—

  (a) direct that the party in default shall not be entitled to rely upon any allegation or material which was the subject matter of the order;
  (b) draw such adverse inferences from the act of non-compliance as the circumstances justify;
  (c) proceed to an award on the basis of such materials as have been properly provided to it;
  (d) make such order as it thinks fit as to the payment of costs of the arbitration incurred in consequence of the non-compliance.

Powers of court in relation to arbitral proceedings

Enforcement of peremptory orders of tribunal.

42.—(1) Unless otherwise agreed by the parties, the court may make an order requiring a party to comply with a peremptory order made by the tribunal.

(2) An application for an order under this section may be made—

  (a) by the tribunal (upon notice to the parties),
  (b) by a party to the arbitral proceedings with the permission of the tribunal (and upon notice to the other parties), or
  (c) where the parties have agreed that the powers of the court under this section shall be available.

(3) The court shall not act unless it is satisfied that the applicant has exhausted any available arbitral process in respect of failure to comply with the tribunal's order.

(4) No order shall be made under this section unless the court is satisfied that the person to whom the tribunal's order was directed has failed to comply with it within the time prescribed in the order or, if no time was prescribed, within a reasonable time.

(5) The leave of the court is required for any appeal from a decision of the court under this section.

Securing the attendance of witnesses.

43.—(1) A party to arbitral proceedings may use the same court procedures as are available in relation to legal proceedings to secure the attendance before the tribunal of a witness in order to give oral testimony or to produce documents or other material evidence.

(2) This may only be done with the permission of the tribunal or the agreement of the other parties.

(3) The court procedures may only be used if —
    (a) the witness is in the United Kingdom, and
    (b) the arbitral proceedings are being conducted in England and Wales or, as the case may be, Northern Ireland.

(4) A person shall not be compelled by virtue of this section to produce any document or other material evidence which he could not be compelled to produce in legal proceedings.

44. — (1) Unless otherwise agreed by the parties, the court has for the purposes of and in relation to arbitral proceedings the same power of making orders about the matters listed below as it has for the purposes of and in relation to legal proceedings.

Court powers exercisable in support of arbitral proceedings.

(2) Those matters are —
    (a) the taking of the evidence of witnesses;
    (b) the preservation of evidence;
    (c) making orders relating to property which is the subject of the proceedings or as to which any question arises in the proceedings —
        (i) for the inspection, photographing, preservation, custody or detention of the property, or
        (ii) ordering that samples be taken from, or any observation be made of or experiment conducted upon, the property;
      and for that purpose authorising any person to enter any premises in the possession or control of a party to the arbitration;
    (d) the sale of any goods the subject of the proceedings;
    (e) the granting of an interim injunction or the appointment of a receiver.

(3) If the case is one of urgency, the court may, on the application of a party or proposed party to the arbitral proceedings, make such orders as it thinks necessary for the purpose of preserving evidence or assets.

(4) If the case is not one of urgency, the court shall act only on the application of a party to the arbitral proceedings (upon notice to the other parties and to the tribunal) made with the permission of the tribunal or the agreement in writing of the other parties.

(5) In any case the court shall act only if or to the extent that the arbitral tribunal, and any arbitral or other institution or person vested by the parties with power in that regard, has no power or is unable for the time being to act effectively.

(6) If the court so orders, an order made by it under this section shall cease to have effect in whole or in part on the order of the tribunal or of any such arbitral or other institution or person having power to act in relation to the subject-matter of the order.

(7) The leave of the court is required for any appeal from a decision of the court under this section.

Determination of preliminary point of law.

45.—(1) Unless otherwise agreed by the parties, the court may on the application of a party to arbitral proceedings (upon notice to the other parties) determine any question of law arising in the course of the proceedings which the court is satisfied substantially affects the rights of one or more of the parties.

An agreement to dispense with reasons for the tribunal's award shall be considered an agreement to exclude the court's jurisdiction under this section.

(2) An application under this section shall not be considered unless—
 (a) it is made with the agreement of all the other parties to the proceedings, or
 (b) it is made with the permission of the tribunal and the court is satisfied—
  (i) that the determination of the question is likely to produce substantial savings in costs, and
  (ii) that the application was made without delay.

(3) The application shall identify the question of law to be determined and, unless made with the agreement of all the other parties to the proceedings, shall state the grounds on which it is said that the question should be decided by the court.

(4) Unless otherwise agreed by the parties, the arbitral tribunal may continue the arbitral proceedings and make an award while an application to the court under this section is pending.

(5) Unless the court gives leave, no appeal lies from a decision of the court whether the conditions specified in subsection (2) are met.

(6) The decision of the court on the question of law shall be treated as a judgment of the court for the purposes of an appeal.

But no appeal lies without the leave of the court which shall not be given unless the court considers that the question is one of general importance, or is one which for some other special reason should be considered by the Court of Appeal.

The award

Rules applicable to substance of dispute.

46.—(1) The arbitral tribunal shall decide the dispute—
 (a) in accordance with the law chosen by the parties as applicable to the substance of the dispute, or
 (b) if the parties so agree, in accordance with such other considerations as are agreed by them or determined by the tribunal.

(2) For this purpose the choice of the laws of a country shall be understood to refer to the substantive laws of that country and not its conflict of laws rules.

(3) If or to the extent that there is no such choice or agreement, the tribunal shall apply the law determined by the conflict of laws rules which it considers applicable.

47.—(1) Unless otherwise agreed by the parties, the tribunal may make more than one award at different times on different aspects of the matters to be determined.

(2) The tribunal may, in particular, make an award relating—
  (a) to an issue affecting the whole claim, or
  (b) to a part only of the claims or cross-claims submitted to it for decision.

(3) If the tribunal does so, it shall specify in its award the issue, or the claim or part of a claim, which is the subject matter of the award.

48.—(1) The parties are free to agree on the powers exercisable by the arbitral tribunal as regards remedies.

Remedies.

(2) Unless otherwise agreed by the parties, the tribunal has the following powers.

(3) The tribunal may make a declaration as to any matter to be determined in the proceedings.

(4) The tribunal may order the payment of a sum of money, in any currency.

(5) The tribunal has the same powers as the court—
  (a) to order a party to do or refrain from doing anything;
  (b) to order specific performance of a contract (other than a contract relating to land);
  (c) to order the rectification, setting aside or cancellation of a deed or other document.

49.—(1) The parties are free to agree on the powers of the tribunal as regards the award of interest.

Interest.

(2) Unless otherwise agreed by the parties the following provisions apply.

(3) The tribunal may award simple or compound interest from such dates, at such rates and with such rests as it considers meets the justice of the case—
  (a) on the whole or part of any amount awarded by the tribunal, in respect of any period up to the date of the award;
  (b) on the whole or part of any amount claimed in the arbitration and outstanding at the commencement of the arbitral proceedings but paid before the award was made, in respect of any period up to the date of payment.

(4) The tribunal may award simple or compound interest from the date of the award (or any later date) until payment, at such rates and with such rests as it considers meets the justice of the case, on the outstanding amount of any award (including any award of interest under subsection (3) and any award as to costs).

(5) References in this section to an amount awarded by the tribunal include an amount payable in consequence of a declaratory award by the tribunal.

(6) The above provisions do not affect any other power of the tribunal to award interest.

Extension of time for making award.

50.—(1) Where the time for making an award is limited by or in pursuance of the arbitration agreement, then, unless otherwise agreed by the parties, the court may in accordance with the following provisions by order extend that time.

(2) An application for an order under this section may be made—
  (a) by the tribunal (upon notice to the parties), or
  (b) by any party to the proceedings (upon notice to the tribunal and the other parties),
but only after exhausting any available arbitral process for obtaining an extension of time.

(3) The court shall only make an order if satisfied that a substantial injustice would otherwise be done.

(4) The court may extend the time for such period and on such terms as it thinks fit, and may do so whether or not the time previously fixed (by or under the agreement or by a previous order) has expired.

(5) The leave of the court is required for any appeal from a decision of the court under this section.

Settlement.

51.—(1) If during arbitral proceedings the parties settle the dispute, the following provisions apply unless otherwise agreed by the parties.

(2) The tribunal shall terminate the substantive proceedings and, if so requested by the parties and not objected to by the tribunal, shall record the settlement in the form of an agreed award.

(3) An agreed award shall state that it is an award of the tribunal and shall have the same status and effect as any other award on the merits of the case.

(4) The following provisions of this Part relating to awards (sections 52 to 58) apply to an agreed award.

(5) Unless the parties have also settled the matter of the payment of the costs of the arbitration, the provisions of this Part relating to costs (sections 59 to 65) continue to apply.

52.—(1) The parties are free to agree on the form of an award.

(2) If or to the extent that there is no such agreement, the following provisions apply.

(3) The award shall be in writing signed by all the arbitrators or all those assenting to the award.

(4) The award shall contain the reasons for the award unless it is an agreed award or the parties have agreed to dispense with reasons.

(5) The award shall state the seat of the arbitration and the date when the award is made.

53. Unless otherwise agreed by the parties, where the seat of the arbitration is in England and Wales or Northern Ireland, any award in the proceedings shall be treated as made there, regardless of where it was signed, despatched or delivered to any of the parties.

54.—(1) Unless otherwise agreed by the parties, the tribunal may decide what is to be taken to be the date on which the award was made.

(2) In the absence of any such decision, the date of the award shall be taken to be the date on which it is signed by the arbitrator or, where more than one arbitrator signs the award, by the last of them.

55.—(1) The parties are free to agree on the requirements as to notification of the award to the parties.

(2) If there is no such agreement, the award shall be notified to the parties by service on them of copies of the award, which shall be done without delay after the award is made.

(3) Nothing in this section affects section 56 (power to withhold award in case of non-payment).

56.—(1) The tribunal may refuse to deliver an award to the parties except upon full payment of the fees and expenses of the arbitrators.

(2) If the tribunal refuses on that ground to deliver an award, a party to the arbitral proceedings may (upon notice to the other parties and the tribunal) apply to the court, which may order that—
  (a) the tribunal shall deliver the award on the payment into court by the applicant of the fees and expenses demanded, or such lesser amount as the court may specify,
  (b) the amount of the fees and expenses properly payable shall be determined by such means and upon such terms as the court may direct, and
  (c) out of the money paid into court there shall be paid out such fees and expenses as may be found to be properly payable and the balance of the money (if any) shall be paid out to the applicant.

Marginal notes:

Form of award.

Place where award treated as made.

Date of award.

Notification of award.

Power to withhold award in case of non-payment.

(3) For this purpose the amount of fees and expenses properly payable is the amount the applicant is liable to pay under section 28 or any agreement relating to the payment of the arbitrators.

(4) No application to the court may be made where there is any available arbitral process for appeal or review of the amount of the fees or expenses demanded.

(5) References in this section to arbitrators include an arbitrator who has ceased to act and an umpire who has not replaced the other arbitrators.

(6) The above provisions of this section also apply in relation to any arbitral or other institution or person vested by the parties with powers in relation to the delivery of the tribunal's award.

As they so apply, the references to the fees and expenses of the arbitrators shall be construed as including the fees and expenses of that institution or person.

(7) The leave of the court is required for any appeal from a decision of the court under this section.

(8) Nothing in this section shall be construed as excluding an application under section 28 where payment has been made to the arbitrators in order to obtain the award.

Correction of award or additional award.

57.—(1) The parties are free to agree on the powers of the tribunal to correct an award or make an additional award.

(2) If or to the extent there is no such agreement, the following provisions apply.

(3) The tribunal may on its own initiative or on the application of a party —
  (a) correct an award so as to remove any clerical mistake or error arising from an accidental slip or omission or clarify or remove any ambiguity in the award, or
  (b) make an additional award in respect of any claim (including a claim for interest or costs) which was presented to the tribunal but was not dealt with in the award.

These powers shall not be exercised without first affording the other parties a reasonable opportunity to make representations to the tribunal.

(4) Any application for the exercise of those powers must be made within 28 days of the date of the award or such longer period as the parties may agree.

(5) Any correction of an award shall be made within 28 days of the date the application was received by the tribunal or, where the correction is made by the tribunal on its own initiative, within 28 days of the date of the award or, in either case, such longer period as the parties may agree.

(6) Any additional award shall be made within 56 days of the date of the original award or such longer period as the parties may agree.

(7) Any correction of an award shall form part of the award.

58. — (1) Unless otherwise agreed by the parties, an award made by the tribunal pursuant to an arbitration agreement is final and binding both on the parties and on any persons claiming through or under them.

Effect of award.

(2) This does not affect the right of a person to challenge the award by any available arbitral process of appeal or review or in accordance with the provisions of this Part.

Costs of the arbitration

59. — (1) References in this Part to the costs of the arbitration are to —
    (a) the arbitrators' fees and expenses,
    (b) the fees and expenses of any arbitral institution concerned, and
    (c) the legal or other costs of the parties.

Costs of the arbitration.

(2) Any such reference includes the costs of or incidental to any proceedings to determine the amount of the recoverable costs of the arbitration (see section 63).

60. An agreement which has the effect that a party is to pay the whole or part of the costs of the arbitration in any event is only valid if made after the dispute in question has arisen.

Agreement to pay costs in any event.

61. — (1) The tribunal may make an award allocating the costs of the arbitration as between the parties, subject to any agreement of the parties.

Award of costs.

(2) Unless the parties otherwise agree, the tribunal shall award costs on the general principle that costs should follow the event except where it appears to the tribunal that in the circumstances this is not appropriate in relation to the whole or part of the costs.

62. Unless the parties otherwise agree, any obligation under an agreement between them as to how the costs of the arbitration are to be borne, or under an award allocating the costs of the arbitration, extends only to such costs as are recoverable.

Effect of agreement or award about costs.

63. — (1) The parties are free to agree what costs of the arbitration are recoverable.

The recoverable costs of the arbitration.

(2) If or to the extent there is no such agreement, the following provisions apply.

(3) The tribunal may determine by award the recoverable costs of the arbitration on such basis as it thinks fit.
    If it does so, it shall specify —

(a) the basis on which it has acted, and

(b) the items of recoverable costs and the amount referable to each.

(4) If the tribunal does not determine the recoverable costs of the arbitration, any party to the arbitral proceedings may apply to the court (upon notice to the other parties) which may —

(a) determine the recoverable costs of the arbitration on such basis as it thinks fit, or

(b) order that they shall be determined by such means and upon such terms as it may specify.

(5) Unless the tribunal or the court determines otherwise —

(a) the recoverable costs of the arbitration shall be determined on the basis that there shall be allowed a reasonable amount in respect of all costs reasonably incurred, and

(b) any doubt as to whether costs were reasonably incurred or were reasonable in amount shall be resolved in favour of the paying party.

(6) The above provisions have effect subject to section 64 (recoverable fees and expenses of arbitrators).

(7) Nothing in this section affects any right of the arbitrators, any expert, legal adviser or assessor appointed by the tribunal, or any arbitral institution, to payment of their fees and expenses.

Recoverable fees and expenses of arbitrators.

64. — (1) Unless otherwise agreed by the parties, the recoverable costs of the arbitration shall include in respect of the fees and expenses of the arbitrators only such reasonable fees and expenses as are appropriate in the circumstances.

(2) If there is any question as to what reasonable fees and expenses are appropriate in the circumstances, and the matter is not already before the court on an application under section 63(4), the court may on the application of any party (upon notice to the other parties) —

(a) determine the matter, or

(b) order that it be determined by such means and upon such terms as the court may specify.

(3) Subsection (1) has effect subject to any order of the court under section 24(4) or 25(3)(b) (order as to entitlement to fees or expenses in case of removal or resignation of arbitrator).

(4) Nothing in this section affects any right of the arbitrator to payment of his fees and expenses.

Power to limit recoverable costs.

65. — (1) Unless otherwise agreed by the parties, the tribunal may direct that the recoverable costs of the arbitration, or of any part of the arbitral proceedings, shall be limited to a specified amount.

(2) Any direction may be made or varied at any stage, but this must be done sufficiently in advance of the incurring of costs to which it relates, or the taking of any steps in the proceedings which may be affected by it, for the limit to be taken into account.

Powers of the court in relation to award

66. — (1) An award made by the tribunal pursuant to an arbitration agreement may, by leave of the court, be enforced in the same manner as a judgment or order of the court to the same effect.

Enforcement of the award.

(2) Where leave is so given, judgment may be entered in terms of the award.

(3) Leave to enforce an award shall not be given where, or to the extent that, the person against whom it is sought to be enforced shows that the tribunal lacked substantive jurisdiction to make the award.

The right to raise such an objection may have been lost (see section 73).

(4) Nothing in this section affects the recognition or enforcement of an award under any other enactment or rule of law, in particular under Part II of the Arbitration Act 1950 (enforcement of awards under Geneva Convention) or the provisions of Part III of this Act relating to the recognition and enforcement of awards under the New York convention or by an action on the award.

67. — (1) A party to arbitral proceedings may (upon notice to the other parties and to the tribunal) apply to the court —

Challenging the award: substantive jurisdiction.

    (a) challenging any award of the arbitral tribunal as to its substantive jurisdiction; or

    (b) for an order declaring an award made by the tribunal on the merits to be of no effect, in whole or in part, because the tribunal did not have substantive jurisdiction.

A party may lose the right to object (see section 73) and the right to apply is subject to the restrictions in section 70(2) and (3).

(2) The arbitral tribunal may continue the arbitral proceedings and make a further award while an application to the court under this section is pending in relation to an award as to jurisdiction.

(3) On an application under this section challenging an award of the arbitral tribunal as to its substantive jurisdiction, the court may by order —

    (a) confirm the award,

    (b) vary the award, or

    (c) set aside the award in whole or in part.

(4) The leave of the court is required for any appeal from a decision of the court under this section.

Challenging the award: serious irregularity.

68.—(1) A party to arbitral proceedings may (upon notice to the other parties and to the tribunal) apply to the court challenging an award in the proceedings on the ground of serious irregularity affecting the tribunal, the proceedings or the award.

A party may lose the right to object (see section 73) and the right to apply is subject to the restrictions in section 70(2) and (3).

(2) Serious irregularity means an irregularity of one or more of the following kinds which the court considers has caused or will cause substantial injustice to the applicant—

(a) failure by the tribunal to comply with section 33 (general duty of tribunal);

(b) the tribunal exceeding its powers (otherwise than by exceeding its substantive jurisdiction: see section 67);

(c) failure by the tribunal to conduct the proceedings in accordance with the procedure agreed by the parties;

(d) failure by the tribunal to deal with all the issues that were put to it;

(e) any arbitral or other institution or person vested by the parties with powers in relation to the proceedings or the award exceeding its powers;

(f) uncertainty or ambiguity as to the effect of the award;

(g) the award being obtained by fraud or the award or the way in which it was procured being contrary to public policy;

(h) failure to comply with the requirements as to the form of the award; or

(i) any irregularity in the conduct of the proceedings or in the award which is admitted by the tribunal or by any arbitral or other institution or person vested by the parties with powers in relation to the proceedings or the award.

(3) If there is shown to be serious irregularity affecting the tribunal, the proceedings or the award, the court may—

(a) remit the award to the tribunal, in whole or in part, for reconsideration,

(b) set the award aside in whole or in part, or

(c) declare the award to be of no effect, in whole or in part.

The court shall not exercise its power to set aside or to declare an award to be of no effect, in whole or in part, unless it is satisfied that it would be inappropriate to remit the matters in question to the tribunal for reconsideration.

(4) The leave of the court is required for any appeal from a decision of the court under this section.

Appeal on point of law.

69.—(1) Unless otherwise agreed by the parties, a party to arbitral proceedings may (upon notice to the other parties and to the tribunal) appeal to the court on a question of law arising out of an award made in the proceedings.

An agreement to dispense with reasons for the tribunal's award shall be considered an agreement to exclude the court's jurisdiction under this section.

(2) An appeal shall not be brought under this section except—
  (a) with the agreement of all the other parties to the proceedings, or
  (b) with the leave of the court.
The right to appeal is also subject to the restrictions in section 70(2) and (3).

(3) Leave to appeal shall be given only if the court is satisfied—
  (a) that the determination of the question will substantially affect the rights of one or more of the parties,
  (b) that the question is one which the tribunal was asked to determine,
  (c) that, on the basis of the findings of fact in the award—
      (i) the decision of the tribunal on the question is obviously wrong, or
      (ii) the question is one of general public importance and the decision of the tribunal is at least open to serious doubt, and
  (d) that, despite the agreement of the parties to resolve the matter by arbitration, it is just and proper in all the circumstances for the court to determine the question.

(4) An application for leave to appeal under this section shall identify the question of law to be determined and state the grounds on which it is alleged that leave to appeal should be granted.

(5) The court shall determine an application for leave to appeal under this section without a hearing unless it appears to the court that a hearing is required.

(6) The leave of the court is required for any appeal from a decision of the court under this section to grant or refuse leave to appeal.

(7) On an appeal under this section the court may by order—
  (a) confirm the award,
  (b) vary the award,
  (c) remit the award to the tribunal, in whole or in part, for reconsideration in the light of the court's determination, or
  (d) set aside the award in whole or in part.
The court shall not exercise its power to set aside an award, in whole or in part, unless it is satisfied that it would be inappropriate to remit the matters in question to the tribunal for reconsideration.

(8) The decision of the court on an appeal under this section shall be treated as a judgment of the court for the purposes of a further appeal.

But no such appeal lies without the leave of the court which shall not be given unless the court considers that the question is one of general importance or is one which for some other special reason should be considered by the Court of Appeal.

Challenge or appeal: supplementary provisions.

70. – (1) The following provisions apply to an application or appeal under section 67, 68 or 69.

(2) An application or appeal may not be brought if the applicant or appellant has not first exhausted –
  (a) any available arbitral process of appeal or review, and
  (b) any available recourse under section 57 (correction of award or additional award).

(3) Any application or appeal must be brought within 28 days of the date of the award or, if there has been any arbitral process of appeal or review, of the date when the applicant or appellant was notified of the result of that process.

(4) If on an application or appeal it appears to the court that the award –
  (a) does not contain the tribunal's reasons, or
  (b) does not set out the tribunal's reasons in sufficient detail to enable the court properly to consider the application or appeal,
the court may order the tribunal to state the reasons for its award in sufficient detail for that purpose.

(5) Where the court makes an order under subsection (4), it may make such further order as it thinks fit with respect to any additional costs of the arbitration resulting from its order.

(6) The court may order the applicant or appellant to provide security for the costs of the application or appeal, and may direct that the application or appeal be dismissed if the order is not complied with.
  The power to order security for costs shall not be exercised on the ground that the applicant or appellant is –
  (a) an individual ordinarily resident outside the United Kingdom, or
  (b) a corporation or association incorporated or formed under the law of a country outside the United Kingdom, or whose central management and control is exercised outside the United Kingdom.

(7) The court may order that any money payable under the award shall be brought into court or otherwise secured pending the determination of the application or appeal, and may direct that the application or appeal be dismissed if the order is not complied with.

(8) The court may grant leave to appeal subject to conditions to the same or similar effect as an order under subsection (6) or (7).
  This does not affect the general discretion of the court to grant leave subject to conditions.

Challenge or appeal: effect of order of court.

71. – (1) The following provisions have effect where the court makes an order under section 67, 68 or 69 with respect to an award.

(2) Where the award is varied, the variation has effect as part of the tribunal's award.

(3) Where the award is remitted to the tribunal, in whole or in part, for reconsideration, the tribunal shall make a fresh award in respect of the matters remitted within three months of the date of the order for remission or such longer or shorter period as the court may direct.

(4) Where the award is set aside or declared to be of no effect, in whole or in part, the court may also order that any provision that an award is a condition precedent to the bringing of legal proceedings in respect of a matter to which the arbitration agreement applies, is of no effect as regards the subject matter of the award or, as the case may be, the relevant part of the award.

Miscellaneous

72.—(1) A person alleged to be a party to arbitral proceedings but who takes no part in the proceedings may question—

    (a) whether there is a valid arbitration agreement,

    (b) whether the tribunal is properly constituted, or

    (c) what matters have been submitted to arbitration in accordance with the arbitration agreement,

by proceedings in the court for a declaration or injunction or other appropriate relief.

Saving for rights of person who takes no part in proceedings.

(2) He also has the same right as a party to the arbitral proceedings to challenge an award—

    (a) by an application under section 67 on the ground of lack of substantive jurisdiction in relation to him, or

    (b) by an application under section 68 on the ground of serious irregularity (within the meaning of that section) affecting him;

and section 70(2) (duty to exhaust arbitral procedures) does not apply in his case.

73.—(1) If a party to arbitral proceedings takes part, or continues to take part, in the proceedings without making, either forthwith or within such time as is allowed by the arbitration agreement or the tribunal or by any provision of this Part, any objection—

    (a) that the tribunal lacks substantive jurisdiction,

    (b) that the proceedings have been improperly conducted,

    (c) that there has been a failure to comply with the arbitration agreement or with any provision of this Part, or

    (d) that there has been any other irregularity affecting the tribunal or the proceedings,

he may not raise that objection later, before the tribunal or the court, unless he shows that, at the time he took part or continued to take part in the proceedings, he did not know and could not with reasonable diligence have discovered the grounds for the objection.

Loss of right to object.

(2) Where the arbitral tribunal rules that it has substantive jurisdiction and a party to arbitral proceedings who could have questioned that ruling—

    (a) by any available arbitral process of appeal or review, or

    (b) by challenging the award,

does not do so, or does not do so within the time allowed by the arbitration agreement or any provision of this Part, he may not object later to the tribunal's substantive jurisdiction on any ground which was the subject of that ruling.

Immunity of arbitral institutions, &c.

74.—(1) An arbitral or other institution or person designated or requested by the parties to appoint or nominate an arbitrator is not liable for anything done or omitted in the discharge or purported discharge of that function unless the act or omission is shown to have been in bad faith.

(2) An arbitral or other institution or person by whom an arbitrator is appointed or nominated is not liable, by reason of having appointed or nominated him, for anything done or omitted by the arbitrator (or his employees or agents) in the discharge or purported discharge of his functions as arbitrator.

(3) The above provisions apply to an employee or agent of an arbitral or other institution or person as they apply to the institution or person himself.

Charge to secure payment of solicitors' costs.

75. The powers of the court to make declarations and orders under section 73 of the Solicitors Act 1974 or Article 71H of the Solicitors (Northern Ireland) Order 1976 (power to charge property recovered in the proceedings with the payment of solicitors' costs) may be exercised in relation to arbitral proceedings as if those proceedings were proceedings in the court.

Supplementary

Service of notices, &c.

76.—(1) The parties are free to agree on the manner of service of any notice or other document required or authorised to be given or served in pursuance of the arbitration agreement or for the purposes of the arbitral proceedings.

(2) If or to the extent that there is no such agreement the following provisions apply.

(3) A notice or other document may be served on a person by any effective means.

(4) If a notice or other document is addressed, pre-paid and delivered by post—

    (a) to the addressee's last known principal residence or, if he is or has been carrying on a trade, profession or business, his last known principal business address, or

    (b) where the addressee is a body corporate, to the body's registered or principal office,

it shall be treated as effectively served.

(5) This section does not apply to the service of documents for the purposes of legal proceedings, for which provision is made by rules of court.

(6) References in this Part to a notice or other document include any form of communication in writing and references to giving or serving a notice or other document shall be construed accordingly.

77.—(1) This section applies where service of a document on a person in the manner agreed by the parties, or in accordance with provisions of section 76 having effect in default of agreement, is not reasonably practicable.

Powers of court in relation to service of documents.

(2) Unless otherwise agreed by the parties, the court may make such order as it thinks fit—
    (a) for service in such manner as the court may direct, or
    (b) dispensing with service of the document.

(3) Any party to the arbitration agreement may apply for an order, but only after exhausting any available arbitral process for resolving the matter.

(4) The leave of the court is required for any appeal from a decision of the court under this section.

78.—(1) The parties are free to agree on the method of reckoning periods of time for the purposes of any provision agreed by them or any provision of this Part having effect in default of such agreement.

Reckoning periods of time.

(2) If or to the extent there is no such agreement, periods of time shall be reckoned in accordance with the following provisions.

(3) Where the act is required to be done within a specified period after or from a specified date, the period begins immediately after that date.

(4) Where the act is required to be done a specified number of clear days after a specified date, at least that number of days must intervene between the day on which the act is done and that date.

(5) Where the period is a period of seven days or less which would include a Saturday, Sunday or a public holiday in the place where anything which has to be done within the period falls to be done, that day shall be excluded.
In relation to England and Wales or Northern Ireland, a "public holiday" means Christmas Day, Good Friday or a day which under the Banking and Financial Dealings Act 1971 is a bank holiday.

79.—(1) Unless the parties otherwise agree, the court may by order extend any time limit agreed by them in relation to any matter relating to the arbitral proceedings or specified in any provision of this Part having effect in default of such agreement.

Power of court to extend time limits relating to arbitral proceedings.

This section does not apply to a time limit to which section 12 applies (power of court to extend time for beginning arbitral proceedings, &c.).

(2) An application for an order may be made—
    (a) by any party to the arbitral proceedings (upon notice to the other parties and to the tribunal), or
    (b) by the arbitral tribunal (upon notice to the parties).

(3) The court shall not exercise its power to extend a time limit unless it is satisfied—
    (a) that any available recourse to the tribunal, or to any arbitral or other institution or person vested by the parties with power in that regard, has first been exhausted, and
    (b) that a substantial injustice would otherwise be done.

(4) The court's power under this section may be exercised whether or not the time has already expired.

(5) An order under this section may be made on such terms as the court thinks fit.

(6) The leave of the court is required for any appeal from a decision of the court under this section.

Notice and other requirements in connection with legal proceedings. 80.—(1) References in this Part to an application, appeal or other step in relation to legal preceedings being taken "upon notice" to the other parties to the arbitral proceedings, or to the tribunal, are to such notice of the originating process as is required by rules of court and do not impose any separate requirement.

(2) Rules of court shall be made—
    (a) requiring such notice to be given as indicated by any provision of this Part, and
    (b) as to the manner, form and content of any such notice.

(3) Subject to any provision made by rules of court, a requirement to give notice to the tribunal of legal proceedings shall be construed—
    (a) if there is more than one arbitrator, as a requirement to give notice to each of them; and
    (b) if the tribunal is not fully constituted, as a requirement to give notice to any arbitrator who has been appointed.

(4) References in this Part to making an application or appeal to the court within a specified period are to the issue within that period of the appropriate originating process in accordance with rules of court.

(5) Where any provision of this Part requires an application or appeal to be made to the court within a specified time, the rules of court relating to the reckoning of periods, the extending or abridging of periods, and the consequences of not taking a step within the period prescribed by the rules, apply in relation to that requirement.

(6) Provision may be made by rules of court amending the provisions of this Part—

    (a) with respect to the time within which any application or appeal to the court must be made,

    (b) so as to keep any provision made by this Part in relation to arbitral proceedings in step with the corresponding provision of rules of court applying in relation to proceedings in the court, or

    (c) so as to keep any provision made by this Part in relation to legal proceedings in step with the corresponding provision of rules of court applying generally in relation to proceedings in the court.

(7) Nothing in this section affects the generality of the power to make rules of court.

81.—(1) Nothing in this Part shall be construed as excluding the operation of any rule of law consistent with the provisions of this Part, in particular, any rule of law as to—   **Saving for certain matters governed by common law.**

    (a) matters which are not capable of settlement by arbitration;

    (b) the effect of an oral arbitration agreement; or

    (c) the refusal of recognition or enforcement of an arbitral award on grounds of public policy.

(2) Nothing in this Act shall be construed as reviving any jurisdiction of the court to set aside or remit an award on the ground of errors of fact or law on the face of the award.

82.—(1) In this Part—   **Minor definitions.**

"arbitrator", unless the context otherwise requires, includes an umpire;

"available arbitral process", in relation to any matter, includes any process of appeal to or review by an arbitral or other institution or person vested by the parties with powers in relation to that matter;

"claimant", unless the context otherwise requires, includes a counter-claimant, and related expressions shall be construed accordingly;

"dispute" includes any difference;

"enactment" includes an enactment contained in Northern Ireland legislation;

"legal proceedings" means civil proceedings in the High Court or a county court;

"peremptory order" means an order made under section 41(5) or made in exercise of any corresponding power conferred by the parties;

"premises" includes land, buildings, moveable structures, vehicles, vessels, aircraft and hovercraft;

"question of law" means—

    (a) for a court in England and Wales, a question of the law of England and Wales, and

    (b) for a court in Northern Ireland, a question of the law of Northern Ireland;

"substantive jurisdiction", in relation to an arbitral tribunal, refers to the matters specified in section 30(1)(a) to (c), and references to the tribunal exceeding its substantive jurisdiction shall be construed accordingly.

(2) References in this Part to a party to an arbitration agreement include any person claiming under or through a party to the agreement.

Index of defined expressions: Part I

83. In this Part the expressions listed below are defined or otherwise explained by the provisions indicated—

agreement, agree and agreed	section 5(1)
agreement in writing	section 5(2) to (5)
arbitration agreement	sections 6 and 5(1)
arbitrator	section 82(1)
available arbitral process	section 82(1)
claimant	section 82(1)
commencement (in relation to arbitral proceedings)	section 14
costs of the arbitration	section 59
the court	section 105
dispute	section 82(1)
enactment	section 82(1)
legal proceedings	section 82(1)
Limitation Acts	section 13(4)
notice (or other document)	section 76(6)
party—	
—in relation to an arbitration agreement	section 82(2)
—where section 106(2) or (3) applies	section 106(4)
peremptory order	section 82(1) (and see section 41(5))
premises	section 82(1)
question of law	section 82(1)
recoverable costs	sections 63 and 64
seat of the arbitration	section 3
serve and service (of notice or other document)	section 76(6)
substantive jurisdiction (in relation to an arbitral tribunal)	section 82(1) (and see section 30(1)(a) to (c))
upon notice (to the parties or the tribunal)	section 80
written and in writing	section 5(6)

Transitional provisions.

84.—(1) The provisions of this Part do not apply to arbitral proceedings commenced before the date on which this Part comes into force.

(2) They apply to arbitral proceedings commenced on or after that date under an arbitration agreement whenever made.

(3) The above provisions have effect subject to any transitional provision made by an order under section 109(2) (power to include transitional provisions in commencement order).

PART II

OTHER PROVISIONS RELATING TO ARBITRATION

Domestic arbitration agreements

85. — (1) In the case of a domestic arbitration agreement the provisions of Part I are modified in accordance with the following sections. **Modification of Part I in relation to domestic arbitration agreement.**

(2) For this purpose a "domestic arbitration agreement" means an arbitration agreement to which none of the parties is —
- (a) an individual who is a national of, or habitually resident in, a state other than the United Kingdom, or
- (b) a body corporate which is incorporated in, or whose central control and management is exercised in, a state other than the United Kingdom,

and under which the seat of the arbitration (if the seat has been designated or determined) is in the United Kingdom.

(3) In subsection (2) "arbitration agreement" and "seat of the arbitration" have the same meaning as in Part I (see sections 3, 5(1) and 6).

86. — (1) In section 9 (stay of legal proceedings), subsection (4) (stay unless the arbitration agreement is null and void, inoperative, or incapable of being performed) does not apply to a domestic arbitration agreement. **Staying of legal proceedings.**

(2) On an application under that section in relation to a domestic arbitration agreement the court shall grant a stay unless satisfied —
- (a) that the arbitration agreement is null and void, inoperative, or incapable of being performed, or
- (b) that there are other sufficient grounds for not requiring the parties to abide by the arbitration agreement.

(3) The court may treat as a sufficient ground under subsection (2)(b) the fact that the applicant is or was at any material time not ready and willing to do all things necessary for the proper conduct of the arbitration or of any other dispute resolution procedures required to be exhausted before resorting to arbitration.

(4) For the puposes of this section the question whether an arbitration agreement is a domestic arbitration agreement shall be determined by reference to the facts at the time the legal proceedings are commenced.

Effectiveness of agreement to exclude court's jurisdiction.

87. — (1) In the case of a domestic arbitration agreement any agreement to exclude the jurisdiction of the court under —

(a) section 45 (determination of preliminary point of law), or

(b) section 69 (challenging the award: appeal on point of law),

is not effective unless entered into after the commencement of the arbitral proceedings in which the question arises or the award is made.

(2) For this purpose the commencement of the arbitral proceedings has the same meaning as in Part I (see section 14).

(3) For the purposes of this section the question whether an arbitration agreement is a domestic arbitration agreement shall be determined by reference to the facts at the time the agreement is entered into.

Power to repeal or amend sections 85 to 87.

88. — (1) The Secretary of State may by order repeal or amend the provisions of sections 85 to 87.

(2) An order under this section may contain such supplementary, incidental and transitional provisions as appear to the Secretary of State to be appropriate.

(3) An order under this section shall be made by statutory instrument and no such order shall be made unless a draft of it has been laid before and approved by a resolution of each House of Parliament.

Consumer arbitration agreements

Application of unfair terms regulations to consumer arbitration agreements.

89. — (1) The following sections extend the application of the Unfair Terms in Consumer Contracts Regulations 1994 in relation to a term which constitutes an arbitration agreement.

For this purpose "arbitration agreement" means an agreement to submit to arbitration present or future disputes or differences (whether or not contractual).

(2) In those sections "the Regulations" means those regulations and includes any regulations amending or replacing those regulations.

(3) Those sections apply whatever the law applicable to the arbitration agreement.

Regulations apply where consumer is a legal person.

90. The Regulations apply where the consumer is a legal person as they apply where the consumer is a natural person.

Arbitration agreement unfair where modest amount sought.

91. — (1) A term which constitutes an arbitration agreement is unfair for the purposes of the Regulations so far as it relates to a claim for a pecuniary remedy which does not exceed the amount specified by order for the purposes of this section.

(2) Orders under this section may make different provision for different cases and for different purposes.

(3) The power to make orders under this section is exercisable —
  (a) for England and Wales, by the Secretary of State with the concurrence of the Lord Chancellor,
  (b) for Scotland, by the Secretary of State with the concurrence of the Lord Advocate, and
  (c) for Northern Ireland, by the Department of Economic Development for Northern Ireland with the concurrence of the Lord Chancellor.

(4) Any such order for England and Wales or Scotland shall be made by statutory instrument which shall be subject to annulment in pursuance of a resolution of either House of Parliament.

(5) Any such order for Northern Ireland shall be a statutory rule for the purposes of the Statutory Rules (Northern Ireland) Order 1979 and shall be subject to negative resolution, within the meaning of section 41(6) of the Interpretation Act (Northern Ireland) 1954.

Small claims arbitration in the county court
92. Nothing in Part I of this Act applies to arbitration under section 64 of the County Courts Act 1984.

Exclusion of Part I in relation to small claims arbitration in the county court.

Appointment of judges as arbitrators
93. — (1) A judge of the Commercial Court or an official referee may, if in all the circumstances he thinks fit, accept appointment as a sole arbitrator or as umpire by or by virtue of an arbitration agreement.

Appointment of judges as arbitrators.

(2) A judge of the Commercial Court shall not do so unless the Lord Chief Justice has informed him that, having regard to the state of business in the High Court and the Crown Court, he can be made available.

(3) An official referee shall not do so unless the Lord Chief Justice has informed his that, having regard to the state of official referees' business, he can be made available.

(4) The fees payable for the services of a judge of the Commercial Court or official referee as arbitrator or umpire shall be taken in the High Court.

(5) In this section —
  "arbitration agreement" has the same meaning as in Part I; and
  "official referee" means a person nominated under section 68(1)(a) of the Supreme Court Act 1981 to deal with official referees' business.

(6) The provisions of Part I of this Act apply to arbitration before a person appointed under this section with the modifications specified in Schedule 2.

Statutory arbitrations

Application of Part I to statutory arbitrations.

94. — (1) The provisions of Part I apply to every arbitration under an enactment (a "statutory arbitration"), whether the enactment was passed or made before or after the commencement of this Act, subject to the adaptations and exclusions specified in sections 95 to 98.

(2) The provisions of Part I do not apply to a statutory arbitration if or to the extent that their application —
  (a) is inconsistent with the provisions of the enactment concerned, with any rules or procedure authorised or recognised by it, or
  (b) is excluded by any other enactment.

(3) In this section and the following provisions of this Part "enactment" —
  (a) in England and Wales, includes an enactment contained in subordinate legislation within the meaning of the Interpretation Act 1978;
  (b) in Northern Ireland, means a statutory provision within the meaning of section 1(f) of the Interpretation Act (Northern Ireland) 1954.

General adaptation of provisions in relation to statutory arbitrations.

95. — (1) The provisions of Part I apply to a statutory arbitration —
  (a) as if the arbitration were pursuant to an arbitration agreement and as if the enactment were that agreement, and
  (b) as if the persons by and against whom a claim subject to arbitration in pursuance of the enactment may be or has been made were parties to that agreement.

(2) Every statutory arbitration shall be taken to have its seat in England and Wales or, as the case may be, in Northern Ireland.

Specific adaptations of provisions in relation to statutory arbitrations.

96. — (1) The following provisions of Part I apply to a statutory arbitration with the following adaptations.

(2) In section 30(1) (competence of tribunal to rule on its own jurisdiction), the reference in paragraph (a) to whether there is a valid arbitration agreement shall be construed as a reference to whether the enactment applies to the dispute or difference in question.

(3) Section 35 (consolidation of proceedings and concurrent hearings) applies only so as to authorise the consolidation of proceedings, or concurrent hearings in proceedings, under the same enactment.

(4) Section 46 (rules applicable to substance of dispute) applies with the omission of subsection (1)(b) (determination in accordance with considerations agreed by parties).

97. The following provisions of Part I do not apply in relation to a statutory arbitration—
  (a) section 8 (whether agreement discharged by death of a party);
  (b) section 12 (power of court to extend agreed time limits);
  (c) sections 9(5), 10(2) and 71(4) (restrictions on effect of provision that award condition precedent to right to bring legal proceedings).

Provisions excluded from applying to statutory arbitrations.

98.—(1) The Secretary of State may make provision by regulations for adapting or excluding any provision of Part I in relation to statutory arbitrations in general or statutory arbitrations of any particular description.

Power to make further provision by regulations.

(2) The power is exercisable whether the enactment concerned is passed or made before or after the commencement of this Act.

(3) Regulations under this section shall be made by statutory instrument which shall be subject to annulment in pursuance of a resolution of either House of Parliament.

PART III
RECOGNITION AND ENFORCEMENT OF CERTAIN FOREIGN AWARDS
Enforcement of Geneva Convention awards

99. Part II of the Arbitration Act 1950 (enforcement of certain foreign awards) continues to apply in relation to foreign awards within the meaning of that Part which are not also New York Convention awards.

Continuation of Part II of the Arbitration Act 1950.

Recognition and enforcement of New York Convention awards

100.—(1) In this Part a "New York Convention award" means an award made, in pursuance of an arbitration agreement, in the territory of a state (other than the United Kingdom) which is a party to the New York Convention.

New York Convention awards.

(2) For the purposes of subsection (1) and of the provisions of this Part relating to such awards—
  (a) "arbitration agreement" means an arbitration agreement in writing, and
  (b) an award shall be treated as made at the seat of the arbitration, regardless of where it was signed, despatched or delivered to any of the parties.
In this subsection "agreement in writing" and "seat of the arbitration" have the same meaning as in Part I.

(3) If Her Majesty by Order in Council declares that a state specified in the Order is a party to the New York Convention, or is a party in respect of any territory so specified, the Order shall, while in force, be conclusive evidence of that fact.

(4) In this section "the New York Convention" means the Convention on the Recognition and Enforcement of Foreign Arbitral Awards adopted

by the United Nations Conference on International Commercial Arbitration on 10th June 1958.

Recognition and enforcement of awards.

101. — (1) A New York Convention award shall be recognised as binding on the persons as between whom it was made, and may accordingly be relied on by those persons by way of defence, set-off or otherwise in any legal proceedings in England and Wales or Northern Ireland.

(2) A New York Convention award may, by leave of the court, be enforced in the same manner as a judgment or order of the court to the same effect.

As to the meaning of "the court" see section 105.

(3) Where leave is so given, judgment may be entered in terms of the award.

Evidence to be produced by party seeking recognition or enforcement.

102. — (1) A party seeking the recognition or enforcement of a New York Convention award must produce —

    (a) the duly authenticated original award or a duly certified copy of it, and

    (b) the original arbitration agreement or a duly certified copy of it.

(2) If the award or agreement is in a foreign language, the party must also produce a translation of it certified by an official or sworn translator or by a diplomatic or consular agent.

Refusal of recognition or enforcement.

103. — (1) Recognition or enforcement of a New York Convention award shall not be refused except in the following cases.

(2) Recognition or enforcement of the award may be refused if the person against whom it is invoked proves —

    (a) that a party to the arbitration agreement was (under the law applicable to him) under some incapacity;

    (b) that the arbitration agreement was not valid under the law to which the parties subjected it or, failing any indication thereon, under the law of the country where the award was made;

    (c) that he was not given proper notice of the appointment of the arbitrator or of the arbitration proceedings or was otherwise unable to present his case;

    (d) that the award deals with a difference not contemplated by or not falling within the terms of the submission to arbitration or contains decisions on matters beyond the scope of the submission to arbitration (but see subsection (4));

    (e) that the composition of the arbitral tribunal or the arbitral procedure was not in accordance with the agreement of the parties or, failing such agreement, with the law of the country in which the arbitration took place;

    (f) that the award has not yet become binding on the parties, or has been set aside or suspended by a competent authority of the country in which, or under the law of which, it was made.

(3) Recognition or enforcement of the award may also be refused if the award as in respect of a matter which is not capable of settlement by arbitration, or if it would be contrary to public policy to recognise or enforce the award.

(4) An award which contains decisions on matters not submitted to arbitration may be recognised or enforced to the extent that it contains decisions on matters submitted to arbitration which can be separated from those on matters not so submitted.

(5) Where an application for the setting aside or suspension of the award has been made to such a competent authority as is mentioned in subsection (2)(f), the court before which the award is sought to be relied upon may, if it considers it proper, adjourn the decision on the recognition or enforcement of the award.

It may also on the application of the party claiming recognition or enforcement of the award order the other party to give suitable security.

104. Nothing in the preceding provisions of this Part affects any right to rely upon or enforce a New York Convention award at common law or under section 66.

Saving for other bases of recognition or enforcement.

PART IV

GENERAL PROVISIONS

105. – (1) In this Act "the court" means the High Court or a county court, subject to the following provisions.

Meaning of "the court": jurisdiction of High Court and county court.

(2) The Lord Chancellor may by order make provision –
    (a) allocating proceedings under this Act to the High Court or to county courts; or
    (b) specifying proceedings under this Act which may be commenced or taken only in the High Court or in a county court.

(3) The Lord Chancellor may by order make provision requiring proceedings of any specified description under this Act in relation to which a county court has jurisdiction to be commenced or taken in one or more specified county courts.

Any jurisdiction so exercisable by a specified county court is exercisable throughout England and Wales or, as the case may be, Northern Ireland.

(4) An order under this section –
    (a) may differentiate between categories of proceedings by reference to such criteria as the Lord Chancellor sees fit to specify, and
    (b) may make such incidental or transitional provision as the Lord Chancellor considers necessary or expedient.

(5) An order under this section for England and Wales shall be made by

statutory instrument which shall be subject to annulment in pursuance of a resolution of either House of Parliament.

(6) An order under this section for Northern Ireland shall be a statutory rule for the purposes of the Statutory Rules (Northern Ireland) Order 1979 which shall be subject to annulment in pursuance of a resolution of either House of Parliament in like manner as a statutory instrument and section 5 of the Statutory Instruments Act 1946 shall apply accordingly.

Crown application.
106. — (1) Part I of this Act applies to any arbitration agreement to which Her Majesty, either in right of the Crown or of the Duchy of Lancaster or otherwise, or the Duke of Cornwall, is a party.

(2) Where Her Majesty is party to an arbitration agreement otherwise than in right of the Crown, Her Majesty shall be represented for the purposes of any arbitral proceedings —

  (a) where the agreement was entered into by Her Majesty in right of the Duchy of Lancaster, by the Chancellor of the Duchy or such person as he may appoint, and

  (b) in any other case, by such person as Her Majesty may appoint in writing under the Royal Sign Manual.

(3) Where the Duke of Cornwall is party to an arbitration agreement, he shall be represented for the purposes of any arbitral proceedings by such person as he may appoint.

(4) References in Part I to a party or the parties to the arbitration agreement or to arbitral proceedings shall be construed, where subsection (2) or (3) applies, as references to the person representing Her Majesty or the Duke of Cornwall.

Consequential amendments and repeals.
107. — (1) The enactments specified in Schedule 3 are amended in accordance with that Schedule, the amendments being consequential on the provisions of this Act.

(2) The enactments specified in Schedule 4 are repealed to the extent specified.

Extent.
108. — (1) The provisions of this Act extend to England and Wales and, except as mentioned below, to Northern Ireland.

(2) The following provisions of Part II do not extend to Northern Ireland —

  section 92 (exclusion of Part I in relation to small claims arbitration in the county court), and

  section 93 and Schedule 2 (appointment of judges as arbitrators).

(3) Sections 89, 90 and 91 (consumer arbitration agreements) extend to Scotland and the provisions of Schedules 3 and 4 (consequential amend-

ments and repeals) extend to Scotland so far as they relate to enactments which so extend, subject as follows.

(4) The repeal of the Arbitration Act 1975 extends only to England and Wales and Northern Ireland.

109.—(1) The provisions of this Act come into force on such day as the Secretary of State may appoint by order made by statutory instrument, and different days may be appointed for different purposes.

Commencement.

(2) An order under subsection (1) may contain such transitional provisions as appear to the Secretary of State to be appropriate.

110. This Act may be cited as the Arbitration Act 1996.

Short title.

1(a) *Mandatory provisions of the 1996 Act*

Provision	Section
Stay of legal proceedings	9,10 & 11
Power of court to extend agreed time limits	12
Application of Limitation Acts	13
Power of court to remove arbitrator	24
Effect of death of arbitrator	26(1)
Liability of parties for fees and expenses of arbitrator	28
Immunity of arbitrator	29
Objection to substantive jurisdiction of tribunal	31
Determination of preliminary point of jurisdiction	32
General duty of tribunal	33
Items to be treated as expenses of arbitrators	37(2)
General duty of parties	40
Securing the attendance of witnesses	43
Power to withhold award in case of non payment	56
Effectiveness of agreement for payment of costs in any event	60
Enforcement of award	66
Challenging the award, substantive jurisdiction and serious irregularity	67 & 68
Supplementary provisions, effect of order of court	70 & 71 so far as they relate to sections 67 & 68
Saving for rights of person who takes no part in proceedings	72
Loss of right to object	73
Immunity of arbitral institutions etc.	74
Charge to secure payment of solicitors' costs	75

1(b)   *Powers of the parties under the 1996 Act*

Power	Section
To designate the seat of the arbitration	3(a)
To agree the application of institutional rules	4(3)
To agree when arbitral proceedings are to be regarded as commenced	14(1)
To agree the number of arbitrators and whether chairman or umpire	15(1)
To agree what is to happen in the event of failure of the procedure for the appointment of the arbitral tribunal	18(1)
To agree the functions of the chairman of the tribunal (where such a post has been agreed)	20(1)
To agree the functions of an umpire (where such a post has been agreed) and in particular whether he is to attend the proceedings and when he is to replace the other arbitrators	21(1)
To agree how the tribunal is to make decisions orders and awards (where two or more arbitrators with no chairman or umpire)	22(1)
To agree in what circumstances the authority of an arbitrator can be revoked	23(1)
To terminate the arbitration agreement	23(4)
To agree with an arbitrator the consequences of his resignation	25(1)
To agree whether and if so how a vacancy is to be filled, whether and if so to what extent the previous proceedings should stand and what effect (if any) the arbitrator ceasing to hold office has on any appointment made by him (alone or jointly)	27(1)
To apply to the court for the adjustment of an arbitrator's fees and expenses	28(2)
To decide any procedural or evidential matter	34(1)

To agree that the arbitral proceedings shall be consolidated with other arbitral proceedings or that concurrent hearings shall be held	35(1)
To agree that the tribunal may not appoint experts, legal advisers or assessors	37(1)
To agree the powers exercisable by the tribunal for the purposes of and in relation to the proceedings	38(1)
To agree that the tribunal shall have power to order on a provisional basis any relief which it would have power to grant in a final award[1]	39(1)
To agree on the powers of the tribunal in case of a party's failure to do something necessary for the proper and expeditious conduct of the arbitration	41(1)
To agree otherwise than that the court may make an order requiring compliance with a peremptory order of the arbitrator	42(1)
To agree the basis for determining the dispute (including the applicable law)	46(1)
To agree the powers exercisable by the arbitral tribunal as regards remedies	48(1)
To agree the powers of the tribunal as regards the award of interest	49(1)
To agree the form of an award	52(1)
To agree that, where the seat of the arbitration is in England and Wales or Northern Ireland, an award may be treated as not being made there	53
To agree that the tribunal may not decide the date to be taken as the date of the award	54(1)
To agree the requirements as to notification of the award to the parties.	55(1)
To apply for the amount of fees and expenses properly payable to the arbitrator to be determined by such means and upon such terms as the court may direct	56(2)(b)

Appendix 1

To agree the powers of the tribunal to correct an award or make an additional award   57(1)

To agreed that an award is not final and binding   58(1)

To agree, but only after a dispute has arisen, that a party is to pay the whole or part of the costs of the arbitration   60

To agree the principles upon which the tribunal may allocate the costs of the arbitration as between the parties   61(2)

To extend an agreement as to how the costs of the arbitration are to be borne beyond the recoverable costs   62

To agree what costs of the arbitration are recoverable   63(1)

To agree that the arbitrator may recover fees and expenses above those that are reasonable   64(1)

To agree otherwise than that the tribunal may limit the recoverable costs of the arbitration   65(1)

To challenge an award for lack of substantive jurisdiction [subject to restrictions, section 73, and right may be lost, section 70(2)&(3)]   67(1)

To apply to the court challenging an award on the grounds of serious irregularity affecting the tribunal   68(1)

To agree to dispense with reasons   69(1)

To agree the manner of service of any notice or other document   76(1)

To agree the method of reckoning periods of time   78(1)

[1] These powers are solely in the parties' gift. They do not fall to the arbitrator in the absence of agreement between the parties.

In the absence of the parties agreeing otherwise the powers of the arbitrator remain as set out in the Act.

1(c) *Duties of the parties under the 1996 Act*

Duty	Section
To appoint an arbitrator on request (if alternative procedure not agreed)	16(4),(5) &(6)
To pay arbitrator's reasonable fees and expenses (joint and several)	28(1)
To pay the arbitrator's reasonable fees and expenses (if any) as are appropriate in the circumstances	28(1)
To do all things necessary for the proper and expeditious conduct of the arbitral proceedings including complying without delay with any determination of the tribunal as to procedural or evidential matters, or with any order or directions of the tribunal	40

1(d) *Powers of the arbitrator under the 1996 Act*

1(d)1 Powers that the arbitrator has under the 1996 Act in any event

Power	Section
Where he resigns; to apply to the court for relief from any resulting liability and for an order in respect of his fees (subject to any agreement with the parties as to the consequences of his resignation)	25(3)
Stay proceedings whilst an application is made to the court to determine a preliminary point of jurisdiction	31(5)
Refuse to deliver an award except on full payment of fees	56(1)
Make an award allocating the costs of the arbitration as between the parties (but any award so made must have regard to any agreement concerning the allocation of costs reached by the parties after the dispute has arisen)	61(1)

Continue the proceedings and make a further award whilst an application to the court is pending in relation to an award as to jurisdiction	67(2)

1(d)2 Powers that the arbitrator has under the 1996 Act unless the parties agree otherwise

Power	Section
Rule on own substantive jurisdiction	30(1)
Continue with the arbitration and make an award whilst an application to the court for the determination of a preliminary point of jurisdiction is pending	32(4)
Decide all procedural and evidential matters	34(1)
Fix and, if thought fit, extend the time within which any directions may be given	34(3)
Appoint experts etc	37(1)
Order a claimant to provide security for the costs of the arbitration	38(3)
Give directions in relation to any property which is the subject of the proceedings including inspection, preservation, retention, the taking of samples and the like	38(4)
Direct that a party or a witness be examined on oath or affirmation and administer the oath or take the affirmation	38(5)
Give directions for the preservation of evidence	38(6)
Make an award dismissing a claim in the case of inordinate or inexcusable delay on the part of a claimant	41(3)
Continue with the proceedings in the absence of a party	41(4)
Make a peremptory order if a party fails to comply with any order or direction	41(5)

Make an award dismissing a claim if a claimant fails to comply with an order to provide security for costs	41(6)

Take the following action if a party fails to comply with a peremptory order other than one relating to security for costs   41(7)
(i) direct that the party in default may not rely upon anything that was the subject of the order
(ii) draw adverse inferences from the non-compliance
(iii) proceed to an award on the basis of such materials as have been properly provided
(iv) make such order for costs incurred in consequence of the non-compliance as he thinks fit

Continue with the arbitration and make an award whilst an application to the court for the determination of a preliminary point of law is pending	45(4)
Make more than one award at different times on different aspects of the matter to be determined	47(1)
Make a declaration as to any matter to be determined	48(3)
Order the payment of a sum of money in any currency	48(4)

Have the same power as the court to: 48(5)
(i) order a party to do or refrain from doing anything
(ii) order specific performance of a contract
(iii) order the rectification, setting aside or cancellation of a deed or other document

Award simple or compound interest on: 49(3)
(i) the whole or part of any amount awarded up to the date of the award
(ii) the whole or part of any amount claimed and outstanding at the commencement of the proceedings but paid before the award was made

Award simple or compound interest from the date of the award on the outstanding amount of the award	49(4)
Decide what is to be taken as the date the award was made	54(1)
Correct an award to remove a clerical mistake	57(3)(a)
Make an additional award in respect of any claim presented to the arbitrator but not dealt with in the award	57(3)(b)
Award costs on the general principle that they should follow the event except where it appears to the arbitrator that this is not appropriate	61(2)
Determine the recoverable costs of the arbitration	63(3)
Direct that the recoverable costs of the arbitration shall not exceed a specified amount	65(1)

1(d)3 Additional powers that may be given by the parties to the arbitrator under the 1996 Act

Power	Section
To designate the seat of the arbitration	3(c)
To order consolidation of proceedings or concurrent hearings	35(1)
Any power that the parties may agree to be appropriate	38(1)
To order on a provisional basis any relief which he would have power to grant in a final award	39(1)
Any power that the parties may agree to be appropriate as regards remedies	48(1)
Any power that the parties may agree to be appropriate as regards interest	49(1)
Any power that the parties may agree to be appropriate as regards the correction of an award or making an additional award	57(1)

1(e) *Duties of the arbitrator under the 1996 Act*

Duty	Section
To appoint a third arbitrator (where appropriate)	16(5)
To appoint an umpire (where appropriate)	16(6)
To properly conduct the proceedings*	24(d)(i)
To use all reasonable dispatch*	24(d)(ii)
To determine whether and if so to what extent the previous proceedings shall stand (where appropriate)	27(4)
Not to act in bad faith*	29(1)
To stay proceedings where application for the determination of a preliminary point of jurisdiction is made to the court (if parties so agree)	31(5)
To act fairly and impartially between the parties*	33(1)(a)
To give each party a reasonable opportunity of putting his case and dealing with that of his opponent*	33(1)(a)
To adopt procedures suitable to the circumstances of the particular case*	33(1)(b)
To avoid unnecessary delay and expense*	33(1)(b)
To give the parties a reasonable opportunity to comment on expert/legal adviser's advice/ opinion/information	37(1)(b)
To take into account any order previously made on a provisional basis	39(3)
Decide the dispute in accordance with the law chosen by the parties as applicable to the substance of the dispute OR if the parties so agree, in accordance with such other considerations as are agreed by them or determined by the tribunal	46(1)(a)
In the absence of agreement under 46(1)(a), apply the law determined by the conflict of law rules which the arbitrator considers applicable	46(3)

To specify the issue which is the subject matter of a partial award	47(3)
To terminate the substantive proceedings in response to a settlement and, if requested by the parties and not objected to by the tribunal, record the settlement in an agreed award	51(3)
Make an award in writing, sign the award, give reasons and state the seat of the arbitration in the award	52(3), (4) & (5)
Serve copies of the award on the parties without delay after the award is made (subject to the right to refuse to deliver an award except on full payment of fees)	55
Deliver an award before payment of fees if payment made into court*	56(2)(a)
To provide the opportunity for representations before correcting an award or making an additional award	57(3)
To determine by award the recoverable costs of the arbitration and specify the basis and the items of recoverable costs and the amount referable to each	63(3)
To make a fresh award within 3 months unless a different period is ordered by the court (where award remitted)	71(3)

Except for those items marked* which are mandatory, the parties may agree a different procedure.

1(f) Powers of the court under the 1996 Act

Power	Section
To grant a stay of legal proceedings	9
To extend the time to begin arbitral proceedings	12
To extend the time bar under the Limitation Acts (in very limited circumstances)	13

To remove a sole arbitrator appointed by one party 17
   after the default of the other party where it is
   agreed that each is to appoint an arbitrator

To act where the appointment procedure has failed 18
   (if there is no agreement to an alternative
   procedure)

To remove an arbitrator 24

To assist an arbitrator who resigns (unless the 25
   parties have agreed the consequences of his
   resignation with the arbitrator)

To determine the substantive jurisdiction of the 32
   arbitrator

To enforce peremptory orders of the arbitrator[1] 42

To issue subpoenas (with the permission of the 43
   tribunal or the agreement of the other party)

To make orders to present evidence and to protect 44
   assets.[1]

To determine a preliminary point of law (restricted 45
   to particular circumstances).[1]

To extend a limited time for making an award 50
   (only after exhausting any available arbitral
   process and where substantial injustice would
   otherwise be done)[1]

To order the arbitrator to deliver his award subject 56
   to payment into court of his fees and expenses.

To determine the arbitrator's reasonable fees.[1] 64

To enforce an award in the same manner as a 66
   judgment of the court.

To declare an award to be of no effect because the 67
   tribunal did not have substantive jurisdiction.[2]

To remit an award to the arbitrator in case of 68
   serious irregularity[2]

To set aside or declare an award to be of no effect 68
   in case of serious irregularity (power only to be
   exercised where court satisfied that it would be
   inappropriate to remit)[2]

To take an appeal on a question of law (subject to considerable restrictions)[1,2]   69

To make a declaration or order to charge property recovered in the proceedings with the payment of solicitor's costs   75

To direct how documents are to be served where the manner of service agreed between the parties is not reasonably practicable.   77

To extend time limits agreed between the parties (subject to restrictions)[1]   79

[1] These powers can be excluded by agreement of the parties
[2] The right of a party to apply to the court for it to exercise these powers may be lost in certain circumstances.

It must also be noted that in most cases a party can only apply to the court for it to exercise its powers on notice to the other party.

Appendix 2
The Construction Industry Model Arbitration Rules February 1998 (CIMAR)

NOTES ISSUED BY THE SOCIETY OF CONSTRUCTION ARBITRATORS

Introduction

1. In response to the Bill which was to become the Arbitration Act 1996, the Society of Construction Arbitrators initiated the production of Model Arbitration rules for adoption by all construction institutions and other bodies having interests in construction arbitration. A series of committees was established under the Chairmanship of Lord Justice Auld, including a plenary group, a steering group and a drafting sub-committee which adopted the acronym CIMAR.

2. In the course of its work, the CIMAR steering group published a framework document with suggested draft rules in September 1996. A full draft of the rules was issued in February 1997 and, after wide consultation, the Rules were published as consultation document in April 1997.

3. After further extensive consultation the Rules were recirculated in draft in October 1997 and offered for formal endorsement to all the relevant construction institutions and bodies. This resulted in requests for a number of additions and amendments, while many bodies were prepared to endorse the Rules as printed. This first edition of the Rules, printed in February 1998, lists the bodies who have endorsed the Rules.

4. The drafting and production of this edition has been undertaken principally by John Uff, FEng QC, Peter Aeberli, RIBA barrister, and Christopher Dancaster, FRICS, Secretary SCA. The Rules will be kept under review and further editions produced by a cross-industry review body which is being established under the auspices of the Society of Construction Arbitrators.

Drafting

5. The Arbitration Act 1996 dictates a radically different approach to arbitration Rules. While the 1950 Act contained only general mea-

sures and required that Rules should be fully drafted, the 1996 Act, contains extensive powers which, in most cases, require contracting in.

6. The drafting team had to decide between incorporation by reference and extensive repetition of sections of the Act within the Rules. The steering group was firmly in favour of the former and decided that, in the interest of efficiency, sections of the Act of immediate relevance should be printed after the Rule in question, with other sections necessary to the working of the Rules being printed as an appendix.

7. Apart from incorporating powers direct from the Act, the Rules have two other purposes:

   (1) to extend or amend the provisions of the Act where necessary; and

   (2) to add a general framework to the specific powers and duties in order to provide guidance to users as well as to arbitrators.

   Objective (2) gives rise to the issue of "user friendliness" which has been much debated. One question was whether the Rules should set out extensive procedures or whether they should be as brief as was consistent with their overall purpose. The approach adopted is essentially one of brevity coupled with clarity, which has generally commanded wide support.

8. The Rules are divided into fourteen sections called Rules with numbering within each Rule running to one decimal place only. In addition, having considered various appendices which might be helpful, the choice has narrowed to two, namely definition of terms (Appendix I) and Sections of the act referred to but not printed in the Rules (Appendix II).

Adoption of CIMAR

9 The importance of having the same Rules adopted by all relevant construction institutions and bodies is generally accepted. A large proportion of construction work now spans more than one professional body and disputes necessarily do likewise. There is no good reason for different arbitration rules to exist within the same industry. Specifically, in the light of Section 86 of the 1996 Act not being brought into force, there is no longer an ability to bring Court proceedings in respect of multi-party disputes. If arbitration is to play a proper role in construction disputes, it is imperative that a workable system of joinder should be created. Common Rules are the only way to achieve this in practice (see Rule 3). There are many other aspects of the Rules where a common approach across the industry is highly desirable (for instance, orders for provisional relief, Rule 10).

10. Where any of the contract producing bodies within the industry considers that individual procedural Rules are required, the Rules make express provision for the incorporating of such Rules, for instance in the form of "advisory or model procedures" under Rule 6. There are other express provisions in CIMAR which invite additional Rules. Conversely, however, some Rules will operate only if they are incorporated as drafted by all the relevant institutions. This applies in the case of joinder under Rule 2, where appointment of a common arbitrator must be considered by the persons individually charged with making the appointment.

Appendix 2

The following tables are intended to draw the reader's attention in summary form to the effect of the adoption of CIMAR as the Rules governing an arbitration upon the powers and duties of those involved.

2(a) *CIMAR and the parties' powers*

Extensions and amendments to the parties' powers

The parties agree by the use of CIMAR (Rule 1.3) that they will not amend the Rules or impose procedures that conflict with them, after an arbitrator is appointed, without the agreement of the arbitrator

Power	Rule number
To apply to a person so empowered for the appointment of an arbitrator	2.3
To give notice of any other dispute falling under the same arbitration agreement. (before an arbitrator has been appointed)	3.2
To give a further notice of arbitration referring any other dispute which falls under the same arbitration agreement to those arbitral proceedings. (after an arbitrator has been appointed)	3.3
To agree with the other party that one single award shall be made if the arbitrator orders concurrent hearings	3.8
To agree that the arbitrator may order consolidation of two or more proceedings involving some common issue	3.9
To agree that more than one award shall be made where the arbitrator has ordered the consolidation of two or more arbitral proceedings	3.10
To request reasons for an order for security for costs. (The arbitrator has discretion to refuse to comply with such a request).	4.7

To agree to extend the one day hearing set for the Short Hearing procedure	7.3
To adduce expert evidence but not recover any costs so incurred unless the arbitrator decides otherwise (in the case of a short hearing procedure)	7.5
To reply to the statement of the other party (in the case of a documents only procedure)	8.3
To be represented in the proceedings by one or more persons of his choice and by different persons at different times	14.1

2(b)   CIMAR *and the parties' duties*

Extensions and amendments to the parties' duties

Duty	**Rule number**
To name any persons proposed as arbitrator	2.2
To provide estimates of the amount in dispute, a view of the need for and the length of any hearing and proposals as to the appropriate form of procedure	6.2
To submit or exchange statements in a particular form	7.2, 8.2, 9.2
To provide formal evidence in relation to an application for provisional relief	10.2
To obtain the permission of the arbitrator before applying to the court requiring compliance with a peremptory order	11.5
To provide formal evidence with an application to dismiss a claim	11.6
To inform the arbitrator of any settlement	14.4
To inform the arbitrator of any intended application to the court and provide copies of any proceedings issued	14.5

2(c)   CIMAR *and the arbitrator's powers*

Extensions and amendments to the arbitrator's powers

Power	Rule number
To decide whether a further dispute should be referred to and consolidated with the proceedings	3.3
To decide or abrogate any matter which may be a condition precedent to the bringing of a further dispute before him	3.5
To order concurrent hearings	3.7
To give such directions as are necessary for hearings ordered under Rule 3.7	3.8
To order that two or more proceedings be consolidated if all parties so agree	3.9
To give such directions as are appropriate for consolidated proceedings ordered under Rule 3.9	3.10
To revoke orders on concurrent hearings or consolidation	3.11
To exercise powers either separately or jointly in consolidated proceedings	3.12
To order the preservation of work, goods or materials that form part of work which is continuing	4.4
To direct the manner in which, by whom and when any test or experiment is carried out	4.5
To observe tests or experiments in the absence of one or both parties	4.5
To give reasons for any decision on security for costs if requested and if he considers it appropriate	4.7
To vary or amend directions given as to procedure	6.3
To dispense with the procedural meeting required by Rule 6.3 if the parties so agree	6.6

To form his own opinion on the matters in dispute and need not inform the parties before delivering his award. (Short Hearing procedure)	7.4
To put questions to or request further amplification from a party (Documents Only procedure)	8.4
To hold a hearing of one day's duration (Documents Only procedure)	8.4
To permit the amendment of statements of claim or defence (Full procedure)	9.3
To require delivery at any time in writing of • an advocate's submission or speech • questions intended to be put to a witness • answers by a witness to identified questions	9.6
To order relief on a provisional basis after application by a party or on his own motion after due notice[1]	10.1, 10.2
To give reasons for a provisional order if considered appropriate	10.3
To order payment to a stakeholder	10.4
Where there is a hearing dealing with part of a dispute: • to decide what issues are to be dealt with • to decide whether on not to give an award • to make an order for relief on a provisional basis	12.2
To supervise or make arrangements for the supervision of work ordered by an award	12.7
To notify an award as a draft or proposal	12.10
To order that the whole or part of an award be paid to a stakeholder pending resolution of a counterclaim	12.11
To order the parties to submit at any time statements of costs incurred and foreseen	13.8

[1] CIMAR give this power, which he would otherwise not have, to the arbitrator.

2(d)   CIMAR *and the arbitrator's duties*

Extensions and amendments to the arbitrator's duties

Duty	Rule number
To deliver separate awards on proceedings for which he has ordered concurrent hearings (unless the parties otherwise agree)	3.8
To deliver a single award on consolidated proceedings (unless the parties otherwise agree)	3.10
To require evidence by affidavit or that some other formal record of the evidence be made (in the case of an application for security for costs, to strike out for want of prosecution, an application for an order for relief on a provisional basis or other instances where he considers it appropriate)	5.6
To convene a procedural meeting and to direct the procedure for the arbitration	6.3
To give directions that he considers appropriate	6.4
To have regard to any advisory procedure and to give effect to any supplementary procedure issued for use under any contract to which the dispute relates	6.5, 13.5
To convene a hearing of not more than one day (Short Hearing procedure)	7.3
To make his award within one month (Short Hearing and Documents Only procedures)	7.6, 8.5
To give detailed directions for all steps in the proceedings (Full Procedure)	9.4
To fix the length of the hearing (Full Procedure)	9.5
To inform the parties after a hearing of a target date for the delivery of his award	12.3
In allocating costs to have regard to all material circumstances as may be relevant	13.2

To deal with the costs of a claim and counterclaim separately unless they are so interconnected that they should be dealt with together	13.3
To take into account claims which are not claims for money when fixing a limit on recoverable costs	13.7
To have regard to any offer of settlement or compromise when allocating costs	13.9
To establish postal addresses, etc.	14.2

2(e)   CIMAR *and persons empowered to appoint arbitrators*

A person empowered to appoint an arbitrator under a contract that incorporates CIMAR has certain duties imposed upon him. These are as follows:

Duty	**Rule number**
To give due consideration as to whether the same or a different arbitrator should be appointed to two or more related arbitral proceedings on the same project	2.6
To appoint the same arbitrator to two or more related arbitral proceedings on the same project unless sufficient grounds shown for not doing so	2.6
To consult with every other person empowered to appoint an arbitrator where different persons are required to appoint in respect of two or more related arbitral proceedings on the same project	2.7
To consider the appointment of an arbitrator already appointed in one arbitral proceeding by another person empowered to appoint to a further arbitral proceeding on the same project	2.7

Permission has been obtained from the copyright holder, the Society of Construction Arbitrators, to reproduce CIMAR in full.

As published, CIMAR include the sections of the Arbitration Act 1996 which are referred to therein. These have not been reproduced here to avoid duplication with Appendix 1.

THE CONSTRUCTION INDUSTRY MODEL ARBITRATION RULES FEBRUARY 1998

FOR USE WITH
ARBITRATION AGREEMENTS UNDER
THE ARBITRATION ACT 1996

PREFACE

The Construction Industry Model Arbitration Rules are the result of extensive consultation with the industry over a period of some eighteen months. This edition of the Rules may be cited as "CIMAR 1998"

At the time of publication endorsement of the use of the Rules has been indicated by the following bodies:

The Association of Consulting Engineers; The British Institute of Architectural Technologists; The British Property Federation; The Chartered Institute of Arbitrators; The Chartered Institute of Building; The Chartered Institution of Building Services Engineers; The Civil Engineering Contractors' Association; Construction Confederation; The Constructors Liaison Group; The Institution of Mechanical Engineers; The Institution of Electrical Engineers; The Royal Institute of British Architects; The Royal Institution of Chartered Surveyors; The Specialist Engineering Contractors' Group.

The Conditions of Contract Standing Joint Committee sponsored by

The Institution of Civil Engineers, The Civil Engineering Contractors Association, The Association of Consulting Engineers has recommended to its sponsors that arbitration under the family of ICE Conditions should offer CIMAR in addition to the ICE Arbitration Procedure as one of its two options

The Joint Contracts Tribunal is publishing an edition of CIMAR incorporating advisory procedures for use with its standard form contracts and is issuing amendments to its contracts to incorporate these Rules.

The Institution of Chemical Engineers will be included a note to their Model Form Contracts that CIMAR may, by agreement between the parties, be used as an alternative to the IChemE arbitration Rules.

ACKNOWLEDGEMENTS

The Society of Construction Arbitrators acknowledges the contribution of all who have assisted the Society in the production of these Rules. In particular the Society acknowledges the involvement of the Chartered Institute of Arbitrators and Miss Elizabeth Dawson. It also acknowledges the financial assistance given by the Royal Institution of Chartered Surveyors.

Sections of the Arbitration Act 1996 are reproduced with the permission of the Controller of Her Majesty's Stationery Office.

SCOTLAND

Users should note that these Rules are not intended for use under Scots law.

DISCLAIMER

There can be no implication of any acceptance of any liability or responsibility whatsoever by the Society of Construction Arbitrators, its officers, members, servants or agents or any other body or person involved in the drafting of these Rules, their servants or agents for the consequences of the use or misuse of these Rules nor for any loss occasioned by any person firm or company acting or refraining from acting as a result of anything contained within these Rules.

THE CONSTRUCTION INDUSTRY
MODEL ARBITRATION RULES

RULE 1: OBJECTIVE AND APPLICATION

1.1 These Rules are to be read consistently with the Arbitration Act 1996 (the Act), with common expressions having the same meaning. Appendix 1 contains definitions of terms. Section numbers given in these Rules are references to the Act.

1.2 The objective of the Rules is to provide for the fair, impartial, speedy, cost-effective and binding resolution of construction disputes, with each party having a reasonable opportunity to put his case and to deal with that of his opponent. The parties and the arbitrator are to do all things necessary to achieve this objective: see Sections 1 (General principles), 33 (General duty of the tribunal) and 40 (General duty of parties)

1.3 After an arbitrator has been appointed under these Rules, the parties may not, without the agreement of the arbitrator, amend the Rules or impose procedures in conflict with them.

1.4 The arbitrator has all the powers and is subject to all the duties under the Act except where expressly modified by the Rules.

1.5 Sections of the Act which need to be read with the Rules are printed with the text. Other Sections referred to are printed in Appendix II.

1.6 These Rules apply where:

(a) a single arbitrator is to be appointed, and

(b) the seat of the arbitration is in England and Wales or Northern Ireland.

1.7 These Rules do not exclude the powers of the Court in respect of arbitral proceedings, nor any agreement between the parties concerning those powers.

RULE 2: BEGINNING AND APPOINTMENT

2.1 Arbitral proceedings are begun in respect of a dispute when one party serves on the other a written notice of arbitration identifying the dispute and requiring him to agree to the appointment of an arbitrator: but see Rule 3.6 and Section 13 (Application of Limitation Acts).

2.2 The party serving notice of arbitration should name any persons he proposes as arbitrator with the notice or separately. The other party should respond and may propose other names.

2.3 If the parties fail to agree on the name of an arbitrator within 14 days (or any agreed extension) after:

(i) the notice of arbitration is served, or

(ii) a previously appointed arbitrator ceases to hold office for any reason,

either party may apply for the appointment of an arbitrator to the person so empowered.

2.4 In the event of a failure in the procedure for the appointment of an arbitrator under Rule 2.3 and in the absence of agreement, Section 18 (Failure of appointment procedure) applies. In this event the court shall seek to achieve the objectives in Rules 2.6 to 2.8.

2.5 The arbitrator's appointment takes effect upon his agreement to act or his appointment under Rule 2.3, whether or not his terms have been accepted.

2.6 Where two or more related arbitral proceedings on the same project fall under separate arbitration agreements (whether or not between the same parties) any person who is required to appoint an arbitrator must give due consideration as to whether

(i) the same arbitrator, or

(ii) a different arbitrator

should be appointed in respect of those arbitral proceedings and should appoint the same arbitrator unless sufficient grounds are shown for not doing so.

2.7 Where different persons are required to appoint an arbitrator in relation to arbitral proceedings covered by Rule 2.6, due consideration includes consulting with every other such person. Where an arbitrator has already been appointed in relation to one such arbitral proceeding, due consideration includes considering the appointment of that arbitrator.

2.8 As between any two or more persons who are required to appoint, the obligation to give due consideration under Rules 2.6 or 2.7 may be discharged by making arrangements for some other person or body to make the appointment in relation to disputes covered by Rule 2.6.

2.9 The provisions in Rules 2 and 3 concerning related arbitral proceedings and disputes and joinder apply in addition to other such provisions contained in any contract between the parties in question.

RULE 3: JOINDER

3.1 A notice of arbitration may include two or more disputes if they fall under the same arbitration agreement.

3.2 A party served with a notice of arbitration may, at any time before an arbitrator is appointed, himself give a notice of arbitration in respect of any other disputes which fall under the same arbitration agreement and those other disputes shall be consolidated with the arbitral proceedings.

3.3 After an arbitrator has been appointed, either party may give a further notice of arbitration to the other and to the arbitrator referring any other dispute which falls under the same arbitration agreement to those arbitral proceedings. If the other party does not consent to the other dispute being so referred, the arbitrator may, as he considers appropriate, order either:

(i) that the other dispute should be referred to and consolidated with the same arbitral proceedings, or
(ii) that the other dispute should not be so referred.

3.4 If the arbitrator makes an order under Rule 3.3(ii), Rules 2.3 and 2.4, then apply.

3.5 In relation to a notice of arbitration in respect of any other dispute under Rules 3.2 or 3.3, the arbitrator is empowered to:

(i) decide any matter which may be a condition precedent to bringing the other dispute before the arbitrator;
(ii) abrogate any condition precedent to the bringing of arbitral proceedings in respect of the other dispute.

3.6 Arbitral proceedings in respect of any other dispute are begun when the notice of arbitration for that other dispute is served: see Section 13 (Application of Limitation Acts).

3.7 Where the same arbitrator is appointed in two or more related arbitral proceedings on the same project each of which involves some common issue, whether or not involving the same parties, the arbitrator may, if he considers it appropriate, order the concurrent hearing of any two or more such proceedings or of any claim or issue arising in such proceedings: see Section 35 and see also Rule 2.9.

3.8 If the arbitrator orders concurrent hearings he may give such other directions as are necessary or desirable for the purpose of such hearings but shall, unless the parties otherwise agree, deliver separate awards in each of such proceedings, see also Rule 2.9.

3.9 Where the same arbitrator is appointed in two or more arbitral proceedings each of which involves some common issue, whether or not involving the same parties, the arbitrator may, if all the parties so agree, order that any two or more such proceedings shall be consolidated.

3.10 If the arbitrator orders the consolidation of two or more arbitral proceedings he may give such other directions as are appropriate for the purpose of such consolidated proceedings and shall, unless the parties otherwise agree, deliver a single award which shall be final and binding on all the parties to the consolidated proceedings.

3.11 Where an arbitrator has ordered concurrent hearings or consolidation under the foregoing rules he may at any time revoke any orders so made and may give such further orders or directions as are appropriate for the separate hearing and determination of the matters in issue.

3.12 Where two or more arbitral proceedings are ordered to be heard concurrently or to be consolidated, the arbitrator may exercise any or all of the powers in these Rules either separately or jointly in relation to the proceedings to which such order relates.

RULE 4: PARTICULAR POWERS

4.1 The arbitrator has the power set out in Section 30 (1) (Competence of the tribunal to rule on its own jurisdiction). This includes power to rule on what matters have been submitted to arbitration.

4.2 The arbitrator has the powers set out in Section 37 (1) (Power to appoint experts, legal advisers or assessors). This includes power to:

(i) appoint experts or legal advisers to report to him and to the parties;
(ii) appoint assessors to assist him on technical matters.

4.3 The arbitrator has the powers set out in Section 38 (4) to (6) (General powers exercisable by the tribunal). This includes power to give directions for:

(a) the inspection, photographing, preservation, custody or detention of property by the arbitrator, an expert or a party;
(b) ordering samples to be taken from, or any observation be made of or experiment conducted upon, property;
(c) a party or witness to be examined on oath or affirmation and to administer any necessary oath or take any necessary affirmation;
(d) the preservation for the purposes of the proceedings of any evidence in the custody or control of a party.

4.4 The arbitrator may order the preservation of any work, goods or materials even though they form part of work which is continuing.

4.5 The arbitrator may direct the manner in which, by whom and when any test or experiment is to be carried out. The arbitrator may himself observe any test or experiment and in the absence of one or both parties provided that they have the opportunity to be present.

4.6 The arbitrator may order a claimant to give security for the whole or part of the costs likely to be incurred by his opponent in defending a claim if satisfied that the claimant is unlikely to be able to pay those costs if the claim is unsuccessful. In exercising this power, the arbitrator shall consider all the circumstances including the strength of the claim and any defence, and the stage at which the application is made. This power is subject to Section 38 (3).

4.7 The arbitrator may give reasons for any decision under Rule 4.6 if the parties so request and the arbitrator considers it appropriate.

4.8 The arbitrator has the power to order a claimant to give security for the arbitrator's costs: see Section 38(3).

4.9 If, without showing sufficient cause, a claimant fails to comply with an order for security for costs under Rule 4.6, the arbitrator may make a peremptory order to the same effect prescribing such time for compliance as he considers appropriate. If the peremptory order is not complied with, the arbitrator may make an award dismissing the claim: see Rules 11.4 and 11.6.

RULE 5: PROCEDURE AND EVIDENCE

5.1 Subject to these Rules, the arbitrator shall decide all procedural and evidential matters including those set out in Section 34 (2) (Procedural and evidential matters), subject to the right of the parties to agree any matter. This includes the power to direct:

(a) when and where any part of the proceedings is to be held;
(b) the languages to be used in the proceedings and whether translations are to be supplied;
(c) the use of written statements and the extent to which they can be later amended.

5.2 The arbitrator shall determine which documents or classes of documents should be disclosed between and produced by the parties and at what stage.

5.3 Whether or not there are oral proceedings the arbitrator may determine the manner in which the parties and their witnesses are to be examined.

5.4 The arbitrator is not bound by the strict rules of evidence and shall determine the admissibility, relevance or weight of any material sought to be tendered on any matters of fact or opinion by any party.

5.5 The arbitrator may himself take the initiative in ascertaining the facts and the law.

5.6 The arbitrator may fix the time within which any order or direction is to be complied with and may extend or reduce the time at any stage.

5.7 In any of the following cases:

(a) an application for security for costs;
(b) an application to strike out for want of prosecution;
(c) an application for an order for provisional relief;
(d) any other instance where he considers it appropriate,

the arbitrator shall require that evidence be put on affidavit or that some other formal record of the evidence be made.

RULE 6: FORM OF PROCEDURE AND DIRECTIONS

6.1 As soon as he is appointed the arbitrator must consider the form of procedure which is most appropriate to the dispute: see Section 33 (General duty of the tribunal).

6.2 For this purpose the parties shall, as soon as practicable after the arbitrator is appointed, provide to each other and to the arbitrator:

(a) a note stating the nature of the dispute with an estimate of the amounts in issue;
(b) a view as to the need for and length of any hearing;
(c) proposals as to the form of procedure appropriate to the dispute.

6.3 The arbitrator shall convene a procedural meeting with the parties or their representatives at which, having regard to any information that may have been submitted under Rule 6.2, he shall give a direction as to the procedure to be followed. The direction may:

(a) adopt the procedures in Rules 7, 8 or 9;
(b) adopt any part of one or more of these procedures;
(c) adopt any other procedure which he considers to be appropriate;
(d) impose time limits

and may be varied or amended by the arbitrator from time to time.

6.4 The arbitrator shall give such directions as he considers appropriate in accordance with the procedure adopted. He shall also give such other directions under these Rules as he considers appropriate: see particularly Rules 4, 5 and 13.4.

6.5 In deciding what directions are appropriate the arbitrator shall have regard to any advisory procedure and give effect to any supplementary procedure issued for use under any contract to which the dispute relates.

6.6 The matters under Rules 6.3 and 6.4 may be dealt with without a meeting if the parties so agree and the arbitrator considers a meeting to be unnecessary.

RULE 7: SHORT HEARING

7.1 This procedure is appropriate where the matters in dispute are to be determined principally by the arbitrator inspecting work, materials, machinery or the like.

7.2 The parties shall, either at the same time or in sequence as the arbitrator may direct, submit written statements of their cases, including any documents and statements of witnesses relied on.

7.3 There shall be a hearing of not more than one day at which each party will have a reasonable opportunity to address the matters in dispute. The arbitrator's inspection may take place before or after the hearing or may be combined with it. The parties may agree to extend the hearing.

7.4 The arbitrator may form his own opinion on the matters in dispute and need not inform the parties of his opinion before delivering his award.

7.5 Either party may adduce expert evidence but may recover any costs so incurred only if the arbitrator decides that such evidence was necessary for coming to his decision.

7.6 The arbitrator shall make his award within one month of the last of the foregoing steps or within such further time as he may require and notify to the parties.

7.7 The recovery of costs is subject to Rule 13.4: see Section 65 (Power to limit recoverable costs).

RULE 8: DOCUMENTS ONLY

8.1 This procedure is appropriate where there is to be no hearing, for instance, because the issues do not require oral evidence, or because the sums in dispute do not warrant the cost of a hearing.

8.2 The parties shall, either at the same time or in sequence as the arbitrator may direct, submit written statements of their cases including:

(a) an account of the relevant facts or opinions relied on;
(b) statements of witnesses concerning those facts or opinions, signed or otherwise confirmed by the witness;
(c) the remedy or relief sought, for instance, a sum of money with interest.

8.3 Each party may submit a statement in reply to that of the other party.

8.4 After reading the parties' written statements, the arbitrator may:

(a) put questions to or request a further written statement from either party;

(b) direct that there be a hearing of not more than one day at which he may put questions to the parties or to any witness. In this event the parties will also have a reasonable opportunity to comment on any additional information given to the arbitrator.

8.5 The arbitrator shall make his award within one month of the last of the foregoing steps, or within such further time as he may require and notify to the parties.

8.6 The recovery of costs is subject to Rule 13.4: see Section 65 (Power to limit recoverable costs).

RULE 9: FULL PROCEDURE

9.1 Where neither the Documents Only nor the Short Procedure is appropriate, the Full Procedure should be adopted, subject to such modification as is appropriate to the particular matters in issue.

9.2 The parties shall exchange statements of claim and defence in accordance with the following guidelines:

(a) each statement should contain the facts and matters of opinion which are intended to be established by evidence and may include a statement of any relevant point of law which will be contended for;

(b) a statement should contain sufficient particulars to enable the other party to answer each allegation without recourse to general denials;

(c) a statement may include or refer to evidence to be adduced if this will assist in defining the issues to be determined;

(d) the reliefs or remedies sought, for instance, specific monetary losses, must be stated in such a way that they can be answered or admitted;

(e) all statements should adopt a common system of numbering or identification of sections to facilitate analysis of issues. Particulars given in schedule form should anticipate the need to incorporate replies.

9.3 The arbitrator may permit or direct the parties at any stage to amend, expand, summarise or reproduce in some other format any of the statements of claim or defence or so as to identify the matters essentially in dispute, including preparing a list of the matters in issue.

9.4 The arbitrator should give detailed directions, with times or dates for all steps in the proceedings including:

(a) further statements or particulars required;

(b) disclosure and production of documents between the parties: see Rule 5.2;

(c) service of statements of witnesses of fact;
(d) the number of experts and service of their reports;
(e) meetings between experts and/or other persons;
(f) arrangements for any hearing.

9.5 The arbitrator should fix the length of each hearing including the time which will be available to each party to present its case and answer that of its opponent.

9.6 The arbitrator may at any time order the following to be delivered to him and to the other party in writing:

(a) any submission or speech by an advocate;
(b) questions intended to be put to any witness;
(c) answers by any witness to identified questions.

RULE 10: PROVISIONAL RELIEF

10.1 The arbitrator has power to order the following relief on a provisional basis: see Section 39 (Power to make provisional awards)

(a) payment of a reasonable proportion of the sum which is likely to be awarded finally in respect of the claims to which the payment relates, after taking account of any defence or counterclaim that may be available;
(b) payment of a sum on account of any costs of the arbitration, including costs relating to an order under this Rule;
(c) any other relief claimed in the arbitral proceedings.

10.2 The arbitrator may exercise the powers under this Rule after application by a party or of his own motion after giving due notice to the parties.

10.3 An order for provisional relief under this Rule must be based on formal evidence: see Rule 5.7. The arbitrator may give such reasons for his order as he thinks appropriate.

10.4 The arbitrator may order any money or property which is the subject of an order for provisional relief to be paid to or delivered to a stakeholder on such terms as he considers appropriate.

10.5 An order for provisional relief is subject to the final adjudication of the arbitrator who makes it, or of any arbitrator who subsequently has jurisdiction over the dispute to which it relates.

RULE 11: DEFAULT POWERS AND SANCTIONS

11.1 The arbitrator has the power set out in Section 41 (3) (Powers of tribunal in case of party's default) to make an award dismissing a claim.

11.2 The arbitrator has the power set out in Section 41 (4) to proceed in the absence of a party or without any written evidence or submission from a party.

11.3 The arbitrator may by any order direct that if a party fails to comply with that order he will:

(a) refuse to allow the party in default to rely on any allegation or material which was the subject of the order;

(b) draw such adverse inferences from the act of non-compliance as the circumstances justify;

and may, if that party fails to comply without showing sufficient cause, refuse to allow such reliance or draw such adverse inferences and may proceed to make an award on the basis of such materials as have been properly provided, and may make any order as to costs in consequence of such non-compliance.

11.4 In addition to his power under Rule 11.3 the arbitrator has the powers set out in Section 41 (5), (6) and (7) (peremptory orders).

11.5 An application to the court for an order requiring a party to comply with a peremptory order may be made only by or with the permission of the arbitrator: see Section 42 (2) (Enforcement of peremptory orders of tribunal).

11.6 An application to dismiss a claim for inordinate and inexcusable delay or failure to comply with a peremptory order to provide security for costs must be based on formal evidence: see Rule 5.7. Where a claim is dismissed on such a ground, the claim shall be barred and may not be rearbitrated.

RULE 12: AWARDS AND REMEDIES

12.1 The arbitrator has the powers set out in Section 47 (Awards on different issues, &c).

12.2 Where the arbitrator directs or the parties agree to a hearing dealing with part of a dispute, then whether or not there is any agreement between the parties as to such matters, the arbitrator may do any of the following:

(a) decide what are the issues or questions to be determined;
(b) decide whether or not to give an award on part of the claims submitted;
(c) make an order for provisional relief: but see Rule 10.2.

12.3 At the conclusion of a hearing, where the arbitrator is to deliver an award he shall inform the parties of the target date for its delivery. The arbitrator must take all possible steps to complete the award by that date and inform the parties of any reason which prevents him doing so. The award shall not deal with the allocation of costs and/or interest unless the parties have been given an opportunity to address these issues.

12.4 An award shall be in writing, dated and signed by the arbitrator. The award must comply with any other requirements of the contract under which it is given. Section 58 (Effect of award) applies but see Rule 10.

12.5 An award should contain sufficient reasons to show why the arbitrator has reached the decisions contained in it unless the parties otherwise agree or the award is agreed.

12.6 The arbitrator has the powers set out in Section 48 (3), (4) and (5) (Remedies).

12.7 Where an award orders that a party should do some act, for instance carry out specified work, the arbitrator has the power to supervise the performance or, if he thinks it appropriate, to appoint (and to reappoint as may be necessary) a suitable person so to supervise and to fix the terms of his engagement and retains all powers necessary to ensure compliance with the award.

12.8 The arbitrator has the powers set out in Section 49 (3) and (4) (Interest). This is in addition to any power to award contractual interest.

12.9 The arbitrator has the powers set out in Section 57 (3) to (6) (Correction of award or additional award) which are to be exercised subject to the time limits stated.

12.10 The arbitrator may notify an award or any part of an award to the parties as a draft or proposal. In such case unless the arbitrator otherwise directs no further evidence shall be admitted and the arbitrator shall consider only such comments of the parties as are notified to him within such time as he may specify and thereafter the arbitrator shall issue the award.

12.11 Where an award is made and there remains outstanding a claim by the other party, the arbitrator may order that the whole or part of the amount of the award be paid to a stakeholder on such terms as he considers appropriate.

RULE 13: COSTS

13.1 The general principle is that costs should be borne by the losing party: see Section 61 (Award of costs). Subject to any agreement between the parties, the arbitrator has the widest discretion in awarding which party should bear what proportion of the costs of the arbitration.

13.2 In allocating costs the arbitrator shall have regard to all material circumstances, including such of the following as may be relevant:

(a) which of the claims has led to the incurring of substantial costs and whether they were successful;
(b) whether any claim which has succeeded was unreasonably exaggerated;

(c) the conduct of the party who succeeded on any claim and any concession made by the other party;

(d) the degree of success of each party.

See also Rule 13.9.

13.3 Where an award deals with both a claim and a counterclaim, the arbitrator should deal with the recovery of costs in relation to each of them separately unless he considers them to be so interconnected that they should be dealt with together.

13.4 The arbitrator may impose a limit on recoverable costs of the arbitration or any part of the proceedings: see Section 65 (Power to limit recoverable costs). In determining such limit the arbitrator shall have regard primarily to the amounts in dispute.

13.5 In determining a limit on recoverable costs the arbitrator shall also have regard to any advisory procedure and give effect to any supplementary procedure issued for use under any contract to which the dispute relates.

13.6 A direction under Rule 13.4 may impose a limit on part of the costs of the arbitration: see Section 59 (Costs of the arbitration).

13.7 Where proceedings include claims which are not claims for money, the arbitrator shall take these into account as he thinks appropriate when fixing a limit under Rule 13.4.

13.8 A direction under Rule 13.4 may be varied at any time, subject to Section 65(2). For this purpose the arbitrator may require the parties to submit at any time statements of costs incurred and foreseen.

13.9 In allocating costs the arbitrator shall have regard to any offer of settlement or compromise from either party, whatever its description or form. The general principle which the arbitrator should follow is that a party who recovers less overall than was offered to him in settlement or compromise should recover the costs which he would otherwise have been entitled to recover only up to the date on which it was reasonable for him to have accepted the offer, and the offeror should recover his costs thereafter.

13.10 Section 63 (3) to (7) (The recoverable costs of the arbitration) applies to the determination of the recoverable costs of the arbitration (determination by the arbitrator or by the court). where the arbitrator is to determine recoverable costs, he may do so on such basis as he thinks fit. Section 59 (Costs of the arbitration) also applies.

RULE 14: MISCELLANEOUS

14.1 A party may be represented in the proceedings by any one or more persons of his choice and by different persons at different times: see Section 36 (Legal or other representation).

14.2 The arbitrator shall establish and record postal addresses and other means, including facsimile or telex, by which communication in writing may be effected for the purposes of the arbitration. Section 76 (3) to (6) (service of notices, &c) shall apply in addition.

14.3 Section 78 (3) to (5) (Reckoning periods of time) apply to the reckoning of periods of time.

14.4 The parties shall promptly inform the arbitrator of any settlement. Section 51 (settlement) then applies.

14.5 The parties shall promptly inform the arbitrator of any intended application to the court and provide copies of any proceedings issued in relation to any such matter.

CIMAR – APPENDIX I

DEFINITION OF TERMS

Act   means the Arbitration Act 1996 (cap 23) including any amendment or reenactment.

claim   includes counterclaim.

claimant   includes counterclaimant.

concurrent hearing   means two or more arbitral proceedings being heard together: see Rules 3.7 and 3.8.

consolidation   means two or more arbitral proceedings being treated as one proceeding: see Rules 3.9 and 3.10.

dispute   includes a difference which is subject to a condition precedent to arbitral proceedings being brought: see Rule 3.5.

notice of arbitration   means the written notice which begins arbitral proceedings: see Rules 2.1 and 3.6.

party   means one of the parties to arbitral proceedings.

provisional order   means an order for provisional relief in accordance with Rule 10.

Rule   refers to a separate section of the Rules or a part.

Section   means a Section of the Act.

CIMAR – APPENDIX II

This sets out certain sections of the Act which are referred to in the text of the Rules but not reproduced with the relevant rule. They are again not reproduced to avoid duplication with Appendix I.

Appendix 3
The Institution of Civil Engineers Arbitration Procedure (1997)

Note: As it has not proved possible to obtain permission to reproduce the Procedure in full each section of this Appendix includes those parts of the Procedure to which it refers. The full Procedure is obtainable from Thomas Telford Publishing.

> 'This Procedure (approved February 1997) has been prepared by The Institution of Civil Engineers principally for use with the ICE family of Conditions of Contract and the NEC family of Documents in England and Wales for arbitrations conducted under the Arbitration Act 1996. It may be suitable for use with other contracts and in other jurisdictions. For the purposes of the ICE family of Conditions of Contract this Procedure shall be deemed to be an amendment or modification to the ICE Procedure (1983).'

3(a) The ICE Procedure and the parties' powers

Extensions and amendments to the parties' powers

Power	Rule number
To apply to the President for the appointment of an arbitrator in the case of failure to agree	4.1
To put forward further disputes to the arbitrator (only before the appointment is completed)	5.1
To apply for leave to appear before the arbitrator on any interlocutory matter	10.2
To be represented by a person who is also a witness	13.2
Agree use of the Short Procedure	15.1
To agree that the Arbitrator may award costs to be paid in equal shares (Short Procedure)	16.1
Require that the Short Procedure no longer be used (party becomes liable for arbitrator's fees and the other party's costs	16.2

Appendix 3

Agree use of the Special Procedure for Experts   17.1

To agree that legal costs may be awarded   18.2
    (special Procedure for Experts)

4.1 If within one calendar month from the service of the Notice to Concur the parties fail to appoint an Arbitrator in accordance with Rule 3 either party may apply to the President to appoint an Arbitrator. Alternatively the parties may agree to apply to the President without a Notice to Concur.

5.1 At any time before the Arbitrator's appointment is completed either party may put forward further disputes or differences to be referred to him. This shall be done by serving upon the other party an additional Notice to Refer in accordance with Rule 2.

10.2 Either party may at any time apply to the Arbitrator for leave to appear before him on any interlocutory matter. The Arbitrator may call a procedural meeting for the purpose or deal with the application in correspondence or otherwise as he thinks fit.

13.2 Any party may be represented by an person including in the case of a company or other legal entity a director officer employee or beneficiary of such company or entity. In particular, a person shall not be prevented from representing a party because he is or may be also a witness in the proceedings. Nothing shall prevent a party from being represented by different persons at different times.

15.1 Where the parties so agree (either of their own motion or at the invitation of the Arbitrator) the arbitration shall be conducted in accordance with the following Short Procedure or any variations thereto which the parties and the Arbitrator so agree.

16.1 Unless the parties otherwise agree the Arbitrator shall have no power to award costs to either party and the Arbitrator's own fees and expenses shall be paid in equal shares by the parties. Where one party has agreed to the Arbitrator's fees and expenses the other party by agreeing to this Short Procedure shall be deemed to have agreed likewise to the Arbitrator's fees and expenses.

Provided always that this Rule shall not apply to any dispute which arises after the Short Procedure has been adopted or imposed by the Contract.

16.2 Either party may at any time before the Arbitrator has made his award under this short Procedure require by written notice served on the Arbitrator and the other party that the arbitration shall cease to be conducted in accordance with this Short Procedure. Save only for Rule 16.3 the Short Procedure shall thereupon no longer apply or bind the parties but any evidence already laid before the Arbitrator shall be

273

admissible in further proceedings as if it had been submitted as part of those proceedings and without further proof.

17.1 Where the parties so agree (either of their own motion or at the invitation of the Arbitrator) the hearing and determination of any issues of fact which depend upon the evidence of Experts shall be conducted in accordance with the following Special Procedure.

18.2 Unless the parties otherwise agree and so notify the Arbitrator neither party shall be entitled to any costs in respect of legal representation assistance or other legal work relating to the hearing and determination of factual issues by this Special Procedure.

3(b)   *The ICE Procedure and the parties' duties*

Extensions and amendments to the parties' duties

Duty	Rule number
Must observe any condition precedent before invoking the arbitration agreement	2.1
Serve a Notice to Refer listing all matters upon which arbitration is desired	2.3
Serve a Notice to Concur in the appointment of an arbitrator and list one or more arbitrators (either party may do this but if not done the arbitration does not proceed)	3.1
Agree to the appointment of a suggested arbitrator or propose alternative(s)	3.2
Must provide a copy of an application for the appointment of an arbitrator to the other party	4.2
To consider (with the other party and the arbitrator): • the Short Procedure and the Special Procedure • documents only • preliminary issues • desirability of expert evidence • disclosure of documents • limiting recoverable costs • excluding the right of appeal	6.4

To serve or deliver a statement of its case in a particular form (limited time in case of Short Procedure)	8.1, 8.3, 15.2, 15.3, 17.2
To notify arbitrator if application made to the court	22.1

2.1 Unless otherwise provided in the Contract a dispute or difference shall be deemed to arise when a claim or assertion made by one party is rejected by the other party and that rejection is not accepted, or no response is received within a period of 28 days. Subject only to the due observance of any condition precedent in the Contract or the arbitration agreement either party may then invoke arbitration by serving a Notice to Refer on the other party.

2.3 The Notice to Refer shall list the matters which the Party serving the Notice to Refer wishes to be referred to arbitration. Nothing stated in the Notice shall restrict that party as to the manner in which it subsequently presents its case.

3.1 At the same time as or after serving the Notice to Refer either party may serve upon the other a Notice to Concur in the appointment of an Arbitrator listing therein the names and addresses of one or more persons it proposes as Arbitrator.

3.2 Within 14 days thereafter the other party shall:

    (a) agree in writing to the appointment of one of the persons listed in the Notice to concur or
    (b) propose in like manner an alternative person or persons.

4.2 The application shall be in writing and shall include:

    (a) a copy of the Notice to Refer;
    (b) a copy of the Notice to Concur or the agreement to dispense with same;
    (c) the names and addresses of all parties to the arbitration;
    (d) a brief statement of the nature and circumstances of the dispute;
    (e) a copy of the arbitration clause in the Contract or of the arbitration agreement;
    (f) the appropriate fee;
    (g) confirmation that any conditions precedent to arbitration contained in the Contract or arbitration agreement have been complied with and
    (h) any other relevant document.

A copy of the application, but not supporting documentation, shall be sent at the same time to the other party.

6.4 The parties and the Arbitrator shall consider whether and to what extent:

275

(a) Part F (Short Procedure) or Part G (Special Procedure for Experts) of these Rules shall apply;

(b) the arbitration should proceed on documents only;

(c) progress may be facilitated and costs saved by determining some of the issues in advance of the main hearing;

(d) evidence of Experts may be necessary, or desirable;

(e) disclosure of documents should be ordered;

(f) there should be a limit put on recoverable costs;

(g) where the Act applies to the Arbitration, the parties should enter into an agreement (if they have not already done so) excluding the right to appeal in accordance with the Act

and in general shall consider such other steps as may achieve the speedy and cost effective resolution of the disputes.

8.1 To the extent that the Arbitrator directs, each party shall prepare in writing and shall serve upon the other party or parties and the Arbitrator a statement of its case comprising:

(a) a summary of that party's case;

(b) a summary of that party's evidence;

(c) a statement or summary of the issues between the parties;

(d) a list and/or a summary of the documents relied upon;

(e) any points of law with references to any authorities relied upon;

(f) a statement or summary of any other matters likely to assist the resolution of the disputes or differences between the parties;

(g) any other document or statement that the Arbitrator considers necessary.

The Arbitrator may order any party to answer the other party's case and to give reasons for any disagreement therewith.

8.3 Statements or answers shall contain sufficient detail for the other party to know the case it has to answer. If sufficient detail is not provided the arbitrator may of his own motion or at the request of the other party order the provision of such further information, clarification or elaboration as the Arbitrator may think fit.

15.2 Within 30 days after the preliminary meeting held under Rule 6.1 the claiming party shall set out it case in the form of a file containing:

(a) a statement as to the orders or awards it seeks;

(b) a statement of its reasons for being entitled to such orders or awards;

(c) copies of any documents on which it relies (including statements) identifying the origin and date of each document

and shall deliver copies of the said file to the other party and to the Arbitrator in such manner and within such time as the Arbitrator may direct.

15.3 The other party shall either at the same time or within 30 days of receipt of the claiming party's statement as the Arbitrator may direct deliver to the claiming party and the Arbitrator its statement in the same form as in Rule 15.2.

17.2 Each party shall set out its case on such issues in the form of a file containing:

(a) a statement of the factual findings it seeks;
(b) a report or statement from and signed by each expert upon whom that party relies;
(c) copies of any other documents referred to in each Expert's report or statement or on which the party relies identifying the origin and date of each document

and shall deliver copies of the said file to the other party and to the Arbitrator in such manner and within such time as the Arbitrator may direct.

22.1 If any party applies to the court for leave to appeal against any award or decision or for any other purpose that party shall forthwith notify the Arbitrator of the application.

The Arbitrator may continue the arbitral proceedings, including making further awards, pending a decision by the court.

3(c)   The ICE Procedure and the arbitrator's powers

Extensions and amendments to the arbitrator's powers

Power	Rule number
To summon the parties to a preliminary meeting	6.1
To require short statements concerning the disputes before a preliminary meeting	6.2
To fix the time within which any directions are to be complied with	7.4
To direct statements in a particular form	8.1
To order a party to answer the other party's case	8.1
To order elaboration of a statement	8.3
Where disputes arising under two or more contracts wholly or mainly with the same subject matter have been referred to one arbitrator he may order the whole or part of the matter shall be heard together[1]	9.1

To give one set of reasons where a number of awards have been made after separate disputes have been heard together	9.2
To call procedural meetings at any time	10.1
Give such directions as he sees fit	10.3
To order the parties to agree facts as facts and figures as figures	12.1
To order the experts to meet and prepare a joint report	12.1
To read the documents to be used before the hearing (He must if the parties so require)	12.2
To hear parties or their witnesses at any time or place and to adjourn the arbitration on application of a party	13.1
To start to hear the arbitration once appointment is complete	13.3
To treat any meeting as a part of the hearing	13.4
To decide the order in which the parties may present their cases and the order in which the issues will be heard and determined	13.5
To order that a submission or speech be put into writing	13.6
To allocate the time available at a hearing between the parties	13.7
To hear or determine issues separately	13.8
To order a party to submit a list of witnesses	14.1
To give leave for the giving of expert evidence (Such evidence not admissible otherwise)	14.2
To order that experts appear separately or concurrently	14.3
To order disclosure or exchange of proofs of evidence relating to factual issues	14.4
To order the preparation and disclosure in advance of a list of points or questions to be put in cross-examination	14.4

To order the putting of a question disclosed in advance that otherwise would not be put	14.5
To disallow costs incurred where a question is put that was not disclosed in advance	14.5
To order that any witness statement or expert's report stand as evidence in chief	14.6
To order a witness or expert to deliver written answers to questions arising out of a statement or report	14.6
To view the site or the works (Short Procedure and Special Procedure for Experts)	15.4, 17.3
To permit or require the submission of further documents or information in writing (Short Procedure)	15.4
To permit or require the preparation or delivery of further files (Short Procedure)	15.4
To vary the time periods in the Short Procedure	15.6
To dispense with a meeting required by rule 15.5[1] (Short Procedure)	15.7
To make a provisional order	19.3
To order that monies the subject of a provisional order be paid to a stakeholder	19.4
To make an award allocating costs in such a manner he considers appropriate	19.7
To direct the basis upon which the costs are to be determined	19.7
To order payment of costs in relation to a provisional order and to order that such costs be paid forthwith	19.7

[1] With the agreement of the parties

6.1 As soon as possible after his appointment the Arbitrator may summon the parties to a preliminary meeting for the purpose of giving such directions about the procedure to be adopted in the arbitration as he considers necessary and to deal with the matters referred to in Rule 6.4.

6.2 The Arbitrator may require the parties to submit to him short statements expressing their perceptions of the disputes or differences.
Such statements shall give sufficient detail of the nature of the issues
to enable the Arbitrator and the parties to discuss procedures
appropriate for their settlement at the preliminary meeting.

8.1 To the extent that the Arbitrator directs, each party shall prepare in
writing and shall serve upon the other party or parties and the
Arbitrator a statement of its case comprising:

(a) a summary of that party's case;
(b) a summary of that party's evidence;
(c) a statement or summary of the issues between the parties;
(d) a list and/or a summary of the documents relied upon;
(e) any points of law with references to any authorities relied upon;
(f) a statement or summary of any other matters likely to assist the
resolution of the disputes or differences between the parties;
(g) any other document or statement that the Arbitrator considers
necessary.

The Arbitrator may order any party to answer the other party's case
and to give reasons for any disagreement therewith.

8.3 Statements or answers shall contain sufficient detail for the other
party to know the case it has to answer. If sufficient detail is not
provided the Arbitrator may of his own motion or at the request of the
other party order the provision of such further information, clarification or elaboration as the Arbitrator may think fit.

9.1 Where disputes or differences have arisen under two or more
contracts each concerned wholly or mainly with the same subject
matter and the resulting arbitrations have been referred to the same
Arbitrator he may with the agreement of all the parties concerned or
upon the application of one of the parties being a party to all the
contracts involved order that the whole or any part of the matters at
issue shall be heard together upon such terms or conditions as the
Arbitrator thinks fit.

9.2 Where an order for concurrent hearings has been made under Rule 9.1
the arbitrator shall nevertheless make separate awards unless all the
parties otherwise agree but the Arbitrator may if he thinks fit prepare
one combined set of reasons to cover all the awards.

10.1 The Arbitrator may at any time call such procedural meetings as he
deems necessary to identify or clarify the issues to be decided and the
procedures to be adopted. For this purpose the Arbitrator may
request particular persons to attend on behalf of the parties.

10.3 At any procedural meeting or otherwise the Arbitrator may give such
directions as he thinks fit for the proper conduct of the arbitration.

12.1 In addition to his other powers the Arbitrator shall also have power to:

(a) order that the parties shall agree facts as facts and figures as figures where possible;

(b) order the parties to prepare an agreed and paginated bundle of all documents relied upon by the parties. The agreed bundle shall thereby be deemed to have been entered in evidence without further proof and without being read out at the hearing. Provided always that either party may at the hearing challenge the admissibility of any document in the agreed bundle;

(c) order that any Experts whose reports have been exchanged should meet and prepare a joint report identifying the points in issue and any other matters covered by their reports upon which they are in agreement and those upon which they disagree, stating the reasons for any disagreement.

12.2 Before the hearing the Arbitrator may and if so requested by the parties shall read the documents to be used at the hearing. For this or any other purpose the Arbitrator may require all such documents to be delivered to him at such time and place as he may specify.

13.1 The Arbitrator may hear the parties their representatives and/or witnesses at any time or place and may adjourn the arbitration for any period on the application of any party or as he thinks fit.

13.3 Nothing in these Rules or in any other rule custom or practice shall prevent the Arbitrator from starting to hear the arbitration once his appointment is completed or at any time thereafter.

13.4 Any meeting with or summons before the Arbitrator at which both parties are represented may if the Arbitrator so directs be treated as part of the hearing.

13.5 At or before the hearing and after hearing representations on behalf of each party the Arbitrator may determine the order in which:

(a) the parties will present their cases;

(b) the order in which the issues will be heard and determined.

13.6 The Arbitrator may order any submission or speech by or on behalf of any party to be put into writing and delivered to him and to the other party. A party so ordered shall be entitled if it so wishes to enlarge upon or vary any such submission orally.

13.7 The Arbitrator may at any time (whether before or after the hearing has commenced) allocate the time available for the hearing between the parties and those representing the parties shall then adhere strictly to that allocation. Should a party's representative fail to complete the presentation of that party's case within the time so allowed further time shall only be afforded at the sole discretion of the arbitrator and upon such conditions as to costs as the Arbitrator may see fit to impose.

13.8 The Arbitrator may on the application of either party or of his own motion hear and determine any issue or issues separately.

14.1 The Arbitrator may order a party to submit in advance of the hearing a list of the witnesses it intends to call. That party shall not thereby be bound to call any witness so listed and may add to the list so submitted at any time.

14.2 No expert evidence shall be admissible except by leave of the Arbitrator. Leave may be given on such terms and conditions as the Arbitrator thinks fit. Unless the Arbitrator otherwise orders such terms shall be deemed to include a requirement that a report from each Expert containing the substance of the evidence to be given shall be served upon the other party within a reasonable time before the hearing.

14.3 The Arbitrator may order that Experts appear before him separately or concurrently at the hearing or otherwise so that he may examine them inquisitorially provided always tat at the conclusion of the questioning by the Arbitrator the parties or its representatives shall have the opportunity to put such further questions to any Expert as they may reasonably require.

14.4 The Arbitrator may order disclosure or exchange of proofs of evidence relating to factual issues. The Arbitrator may also order any party to prepare and disclose in writing in advance a list of points or questions to be put in cross-examination of any witness.

14.5 Where a list of questions is disclosed whether pursuant to an order of the Arbitrator or otherwise the party making disclosure shall not be bound to put any question therein to the witness unless the Arbitrator so orders. Where the party making disclosure puts a question not so listed in cross-examination the Arbitrator may disallow the costs thereby occasioned.

14.6 The Arbitrator my order that any witness statement or Expert's report which has been disclosed shall stand as the evidence in chief of that witness or Expert. The Arbitrator may also at any time before cross-examination order the witness or Expert to deliver written answers to questions arising out of any statement or report.

15.4 The Arbitrator may view the site or the works and may in his sole discretion order, permit or require either or both parties to:

(a) submit further documents or information in writing;
(b) prepare or deliver further files by way of reply or response. Such files may include witness statements or expert reports.

17.3 After reading the parties cases the Arbitrator may view the site or the works and may require either or both parties to submit further documents or information in writing.

15.6 The time periods in Rules 15.2, 15.3 and 15.5 may be varied as the Arbitrator may think fit.

15.7 Alternatively with the agreement of the parties the Arbitrator may dispense with the meeting and upon receipt of any further files or information under Rule 15.4 proceed directly to the award in accordance with Rule 15.8.

19.3 The Arbitrator may also make a provisional order and for this purpose the Arbitrator shall have power to award payment by one party to another of a sum representing a reasonable proportion of the final net amount which in his opinion that party is likely to be ordered to pay after determination of all the issues in the arbitration and after taking into account any defence or counterclaim upon which the other party may be entitled to rely.

19.4 The Arbitrator shall have power to order the party against whom a provisional order is made to pay part or all of the sum awarded to a stakeholder. In default of compliance with such an order the Arbitrator may order payment of the whole sum in the provisional order to the other party.

19.7 Unless otherwise provided in this Procedure, the Arbitrator shall have power to:

(a) make an award allocating the Costs of the Arbitration between the parties in such manner as he considers appropriate;
(b) direct the basis upon which the costs are to be determined;
(c) in default of agreement by the parties, determine the amount of the recoverable costs;
(d) order payment of costs in relation to a provisional order including power to order that such costs shall be paid forthwith.

3(d)   *The ICE Procedure and the arbitrator's duties*

Extensions and amendments to the arbitrator's duties

Duty	Rule number
Upon acceptance of an appointment to notify both parties simultaneously in writing	4.1
To direct that money ordered as security shall be paid into a stakeholder account	7.5

To make separate awards where concurrent hearings have been ordered[1]	9.2
To fix a day to meet the parties (Short Procedure) or Experts (Special Procedure for Experts)	15.5, 17.4
To give notice of any particular person to be questioned by him at the meeting (Short Procedure)	15.6
To make an award within 30 days (Short Procedure)	15.8
To inform the parties in writing that his award is made and how and where it may be taken up	21.2

[1] Unless agreed otherwise by the parties

4.1 If within one calendar month from the service of the Notice to Concur the parties fail to appoint an Arbitrator in accordance with Rule 3 either party may apply to the President to appoint an Arbitrator. Alternatively the parties may agree to apply to the President without a Notice to Concur.

7.5 The Arbitrator shall have power to:

(a) make an order for security for costs in favour of one or more of the parties and
(b) order his own costs to be secured.

Money ordered to be paid under this Rule shall be paid as directed by the Arbitrator into a separate bank account in the name of a stake-holder to be appointed by and subject to the directions of the Arbitrator.

9.2 Where an order for concurrent hearings has been made under Rule 9.1 the Arbitrator shall nevertheless make separate awards unless all the parties otherwise agree but the Arbitrator may if he thinks fit prepare one combined set of reasons to cover all the awards.

15.5 Within 30 days of completing the foregoing steps the Arbitrator shall fix a day to meet the parties for the purpose of:

(a) receiving any oral submissions which either party may wish to make;
(b) the Arbitrator putting questions to the parties their representatives or witnesses.

For this purpose the Arbitrator shall give notice of any particular person he wishes to question but no person shall be bound to appear before him.

17.4 Thereafter the Arbitrator shall fix a day when he shall meet the Experts whose reports or statements have been submitted. At the meeting each Expert may address the Arbitrator and put questions to any other Expert representing the other party. The Arbitrator shall so direct the meeting as to ensure that each Expert has an adequate opportunity to explain his opinion and to comment upon any opposing opinion. No other person shall be entitled to address the Arbitrator or question any Expert unless the parties and the Arbitrator so agree.

15.6 The time periods in Rules 15.2, 15.3 and 15.5 may be varied as the Arbitrator may think fit.

15.8 Within 30 days following the conclusion of the meeting under Rule 15.5, or in the absence of a meeting 30 days following receipt of the further files or information under Rule 15.4, or such further period as the Arbitrator may reasonably require the Arbitrator shall make his award.

21.2 When the Arbitrator has made his award (including a provisional order under Rule 19.3) he shall so inform the parties in writing and shall specify how and where it may be taken up upon full payment of his fees and expenses.

Permission to reproduce material from the ICE Arbitration Procedure 1997 has been granted by the copyright holder, the Institution of Civil Engineers. Copies of the Procedure can be obtained from Thomas Telford Publishing, 1 Heron Quay, London E14 4JD.

Appendix 4
Dispute resolution clauses in Standard Form Contracts

4.1 The JCT Forms of Contract

The Joint Contracts Tribunal issued Amendment 18 to the Standard Form of Building Contract 1980 Edition ('JCT 80') in April 1998 in time for the coming into force of the adjudication provisions of the Housing Grants, Construction and Regeneration Act 1996 on 1 May 1998.

These new provisions will generally apply *mutatis mutandis* in all the JCT family of main contracts and in all related sub-contracts. However the 'litigation option' (Article 7A and clause 41C) will, for the time being, only apply to JCT 80, WCD 81 and the Management Form, although it is thought likely that it will be extended to all the other forms in time.

The following is the relevant text in Amendment 18.

Dispute or difference – adjudication

Article 5
If any dispute or difference arises *under* this Contract either Party may refer it to adjudication in accordance with clause 41A.

Dispute or difference – arbitration

Article 7A
Where the entry in the Appendix stating that "clause 41B applies" has not been deleted then, subject to Article 5, if any dispute or difference as to any matter or thing of whatsoever nature arising under this Contract or in connection therewith, except in connection with the enforcement of any decision of an Adjudicator appointed to determine a dispute or difference arising thereunder, shall arise between the Parties either during the progress or after the completion or abandonment of the Works or after the determination of the employment of the Contractor except under clause 31 (*statutory tax deduction scheme*) to the extent provided in clause 31.9, or under clause 3 of the VAT Agreement it shall be referred to arbitration in accordance with clause 41B and the JCT 1998 edition of the Construction Industry Model Arbitration Rules (CIMAR).[g.3]

Article 7B

Where the entry in the Appendix stating that "clause 41B applies" has been deleted then, subject to Article 5, if any dispute or difference as to any matter or thing of whatsoever nature arising under this Contract or in connection therewith shall arise between the Parties either during the progress or after the completion or abandonment of the Works or after the determination of the employment of the Contractor it shall be determined by legal proceedings and clause 41C shall apply to such proceedings.

[**g.3**] The JCT 1998 edition of the Construction Industry Model Arbitration Rules (CIMAR) contain procedures for beginning an arbitration and the appointment of an arbitrator, the consolidation or joinder of disputes including related disputes between different parties engaged under different contracts on the same project, and for the conduct of arbitral proceedings. The objective of CIMAR is the fair, impartial, speedy, cost effective and binding resolution of construction disputes. The JCT 1998 edition of the Construction Industry Model Arbitration Rules (CIMAR) includes additional rules concerning the calling of preliminary meetings and supplemental and advisory procedures which may, with the agreement of the parties, be used with Rules 7 (short hearing), 8 (documents only) or 9 (full procedure).

Part 4: Settlement of disputes – Adjudication – Arbitration – Legal Proceedings

41A Adjudication

41A·1 Clause 41A applies where, pursuant to Article 5, either Party refers any dispute or difference arising under this Contract to adjudication. **Application of clause 41A**

41A·2 The Adjudicator to decide the dispute or difference shall be either an individual agreed by the Parties or, on the application of either Party, an individual to be nominated as the Adjudicator by the person named in the Appendix ("the nominator"). Provided that [**y.1**] **Identity of Adjudicator**

41A·2 ·1 no Adjudicator shall be agreed or nominated under clause 41A.2.2 or clause 41A.3 who will not execute the Standard Agreement for the appointment of an Adjudicator issued by the Joint Contracts Tribunal (the "JCT Adjudication Agreement"†) with the Parties, [**y.1**] and

41A·2 ·2 where either Party has given notice of his intention to refer a dispute to adjudication then

– any agreement by the Parties on the appointment of an Adjudicator must be reached with the object of securing the appointment of, and the referral of the dispute or difference to, the Adjudicator within 7 days of the date of the notice of intention to refer (*see clause 41A.4.1*);

– any application to the nominator must be made with the object of securing the appointment of, and the referral of the dispute or difference to, the Adjudicator within 7 days of the date of the notice of intention to refer;

41A·2 ·3 upon agreement by the Parties on the appointment of the Adjudicator or upon receipt by the Parties from the nominator of the name of the nominated Adjudicator the Parties shall thereupon execute with the Adjudicator the JCT Adjudication Agreement.

[y·1] The nominators named in the Appendix have agreed with the Joint Contracts Tribunal that they will comply with the requirements of clause 41A on the nomination of an adjudicator including the requirement in clause 41A·2·2 for the nomination to be made with the object of securing the appointment of, and the referral of the dispute or difference to, the Adjudicator within 7 days of the date of the notice of intention to refer; and will only nominate adjudicators who will enter into the "JCT Adjudication Agreement".

† The JCT Adjudication Agreement, whose text is set out in the Guidance Notes to this Amendment, is available from the retailers of JCT Forms.

Death of Adjudicator – inability to adjudicate

41A·3 If the Adjudicator dies or becomes ill or is unavailable for some other cause and is thus unable to adjudicate on a dispute or difference referred to him, the Parties may either agree upon an individual to replace the Adjudicator or either party may apply to the nominator for the nomination of an adjudicator to adjudicate that dispute or difference; and the Parties shall execute the JCT Adjudication Agreement with the agreed or nominated Adjudicator.

Dispute or difference – notice of intention to refer to Adjudication – referral

41A·4 ·1 When pursuant to Article 5 a Party requires a dispute or difference to be referred to adjudication then that Party shall give notice to the other Party of his intention to refer the dispute or difference, briefly identified in the notice, to adjudication. Within 7 days from the date of such notice or the execution of the JCT Adjudication Agreement by the Adjudicator if later, the Party giving the notice of intention shall refer the dispute or difference to the Adjudicator for his decision ("the referral"); and shall include with that referral particulars of the dispute or difference together with a summary of the contentions on which he relies, a statement of the relief or remedy which is sought and any material he wishes the Adjudicator to consider. The referral and its accompanying documentation shall be copied simultaneously to the other Party.

41A·4 ·2 The referral by a Party with its accompanying documentation to the Adjudicator and the copies thereof to be provided to the other Party shall be given by actual delivery or by FAX or by registered post or recorded delivery. If given by FAX then, for record purposes, the referral and its accompanying documentation must forthwith be sent by first class post or given by actual delivery. If sent by registered post or recorded delivery the referral and its accompanying documentation shall, subject to proof to the contrary, be deemed to have been received 48 hours after the date of posting subject to the exclusion of Sundays and any Public Holiday.

41A·5 ·1 The Adjudicator shall immediately upon receipt of the referral and its accompanying documentation confirm the date of that receipt to the Parties.

Conduct of the Adjudication

41A·5 ·2 The Party not making the referral may, by the same means stated in clause 41A·4·2, send to the Adjudicator within 7 days of the date of the referral with a copy to the other Party, a written statement of the contentions on which he relies and any material he wishes the Adjudicator to consider.

41A·5 ·3 The Adjudicator shall within 28 days of his receipt of the referral and its accompanying documentation under clause 41A·4·1 and acting as an adjudicator for the purposes of S.108 of the Housing Grants, Construction and Regeneration Act 1996 and not as an expert or an arbitrator reach his decision and forthwith send that decision in writing to the Parties. Provided that the Party who has made the referral may consent to allowing the Adjudicator to extend the period of 28 days by up to 14 days; and that by agreement between the Parties after the referral has been made a longer period than 28 days may be notified jointly by the Parties to the Adjudicator within which to reach his decision.

41A·5 ·4 The Adjudicator shall not be obliged to give reasons for his decision.

41A·5 ·5 In reaching his decision the Adjudicator shall act impartially, set his own procedure and at his absolute discretion may take the initiative in ascertaining the facts and the law as he considers necessary in respect of the referral which may include the following:

·5 ·1 using his own knowledge and/or experience;

·5 ·2 opening up, reviewing and revising any certificate, opinion, decision, requirement or notice issued given or made under the Contract as if no such certificate, opinion, decision, requirement or notice had been issued given or made;

·5 ·3 requiring from the Parties further information than that contained in the notice of referral and its accompanying documentation or in any written statement provided by the Parties including the results of any test that have been made or of any opening up;

·5 ·4 requiring the Parties to carry out tests or additional tests or to open up work or further open up work;

·5 ·5 visiting the site of the Works or any workshop where work is being or has been prepared for the Contract;

·5 ·6 obtaining such information as he considers necessary from any employee or representative of the Parties provided that before obtaining information from an employee of a Party he has given prior notice to that Party;

·5 ·7 obtaining from others such information and advice as he considers necessary on technical and on legal matters subject to giving prior notice to the Parties together with a statement or estimate of the cost involved;

·5 ·8 having regard to any term of the contract relating to the payment of interest deciding the circumstances in which or the period for which a simple rate of interest shall be paid.

41A·5 ·6 Any failure by either Party to enter into the JCT Adjudication Agreement or to comply with any requirement of the Adjudicator under clause 41A·5·5 or with any provision in or requirement under clause 41A shall not invalidate the decision of the Adjudicator.

41A·5 ·7 The Parties shall meet their own costs of the Adjudication except that the Adjudicator may direct as to who should pay the cost of any test or opening up if required pursuant to clause 41A·5·5·4.

Adjudicator's fee and reasonable expenses – payment

41A·6 ·1 The Adjudicator in his decision shall state how payment of his fee and reasonable expenses is to be apportioned as between the Parties. In default of such statement the Parties shall bear the cost of the Adjudicator's fee and reasonable expenses in equal proportions.

41A·6 ·2 The Parties shall be jointly and severally liable to the Adjudicator for his fee and for all expenses reasonably incurred by the Adjudicator pursuant to the Adjudication.

41A·7 ·1   The decision of the Adjudicator shall be binding on the Parties until the dispute or difference is finally determined by arbitration or by legal proceedings or by an agreement in writing between the Parties made after the decision of the Adjudicator has been given. [**y.2**]

Effect of Adjudicator's decision

41A·7 ·2   The Parties shall, without prejudice to their other rights under the Contract, comply with the decisions of the Adjudicator; and the Employer and the Contractor shall ensure that the decisions of the Adjudicator are given effect.

41A·7 ·3   If either Party does not comply with the decision of the Adjudicator the other Party shall be entitled to take legal proceedings to secure such compliance pending any final determination of the referred dispute or difference pursuant to clause 41A·7·1.

> [**y.2**] The arbitration or legal proceedings are **not** an appeal against the decision of the Adjudicator but are a consideration of the dispute or difference as if no decision had been made by an Adjudicator.

41A·8   The Adjudicator shall not be liable for anything done or omitted in the discharge or purported discharge of his functions as Adjudicator unless the act or omission is in bad faith and this protection from liability shall similarly extend to any employee or agent of the Adjudicator.

Immunity

41B   Arbitration

A reference in clause 41B to a Rule or Rules is a reference to the JCT 1998 edition of the Construction Industry Model Arbitration Rules (CIMAR) current at the Base Date.

41B·1 ·1   Where pursuant to Article 7A either Party requires a dispute or difference to be referred to arbitration then that Party shall serve on the other Party a notice of arbitration to such effect in accordance with Rule 2.1 which states:

> "Arbitral proceedings are begun in respect of a dispute when one party serves on the other a written notice of arbitration identifying the dispute and requiring him to agree to the appointment of an arbitrator;"

and an arbitrator shall be an individual agreed by the parties or appointed by the person named in the Appendix in accordance with Rule 2.3 which states:

> "If the parties fail to agree on the name of an arbitrator within 14 days (or any agreed extension) after:
> (i)   the notice of arbitration is served, or
> (ii) a previously appointed arbitrator ceases to hold office for

> any reason either party may apply for the appointment
> of an arbitrator to the person so empowered."

By Rule 2.5:

> "the arbitrator's appointment takes effect upon his agreement
> to act or his appointment under Rule 2.3, whether or not his
> terms have been accepted."

41B·1 ·2   Where two or more related arbitral proceedings in respect of the
Works fall under separate arbitration agreements Rules 2.6, 2.7
and 2.8 shall apply thereto.

41B·1 ·3   After an arbitrator has been appointed either Party may give a
further notice of arbitration to the other Party and to the Arbitrator referring any other dispute which falls under Article 7A to
be decided in the arbitral proceedings and Rule 3.3 shall apply
thereto.

41B·2   Subject to the provisions of Article 7A and clause 30·9 the Arbitrator shall, without prejudice to the generality of his powers,
have power to rectify this Contract so that it accurately reflects the
true agreement made by the Parties, to direct such measurements
and/or valuations as may in his opinion be desirable in order to
determine the rights of the Parties and to ascertain and award any
sum which ought to have been the subject of or included in any
certificate and to open up, review and revise any certificate, opinion, decision, requirement or notice and to determine all mattes
in dispute which shall be submitted to him in the same manner as
if no such certificate, opinion, decision, requirement or notice had
been given.

41B·3   Subject to clause 41B·4 the award of such Arbitrator shall be final
and binding on the Parties.

41B·4   The Parties hereby agree pursuant to section 45(2)(a) and Section
69(2)(a) of the Arbitration Act, 1996, that either Party may (upon
notice to the other Party and to the Arbitrator):

41B·4 ·1   apply to the courts to determine any question of law arising in
the course of the reference; and

41B·4 ·2   appeal to the courts on any question of law arising out of an
award made in an arbitration under this Arbitration Agreement.

41B·5   The provisions of the Arbitration Act 1996 or any amendment
thereof shall apply to any arbitration under this Contract wherever
the same, or any part of it, shall be conducted. [**y.3**]

41B·6   The arbitration shall be conducted in accordance with the JCT 1998
edition of the Construction Industry Model Arbitration Rules

(CIMAR) current at the Base Date. Provided that if any amendments to the Rules so current have been issued by the Joint Contracts Tribunal after the Base Date the Parties may, by a joint notice in writing to the Arbitrator, state that they wish the arbitration to be conducted in accordance with the Rules as so amended.

> **[y.3]** It should be noted that the provisions of the Arbitration Act 1996 do not extend to Scotland. Where the site of the Works is situated in Scotland then the forms issued by the Scottish Building Contract Committee which contain Scots proper law, adjudication and arbitration provisions are the appropriate documents. The SBCC issues guidance in this respect.

41C   Legal Proceedings

41C·1 When any dispute or difference is to be determined by legal proceedings, then insofar as the Conditions provide for the issue of a certificate, or the expression of an opinion or the giving or a decision, requirement or notice such provision shall not prevent the Court, in determining the rights and liabilities of the Parties hereto, from making any finding necessary to establish whether such certificate was correctly issued or opinion correctly expressed or decision, requirement or notice correctly given on the facts found by the Court; nor shall such provision prevent the Court establishing what certificate ought to have been issued or what other opinion should have been expressed or what other decision, requirement or notice should have been given as if no certificate, opinion, decision, requirement or notice had been issued, expressed or given.

The copyright holder for the JCT forms is RIBA Publications Ltd.

While this is not the place for a detailed examination of the provisions for adjudication, in the commentary which follows reference is made to some aspects which impinge upon the provisions for arbitration and litigation.

Article 5

Note that only disputes arising *under* the contract are referable to adjudication. This means that certain types of dispute which may be referred to arbitration under Article 7A as arising out of the contract rather than under it, such as disputes relating to the existence of the contract or to misrepresentation, are not referable to adjudication (unless, of course, both parties agree).

Article 7A

Under this Article all disputes arising under or in connection with the contract are referable to arbitration. The omission of the words 'as to the construction of this contract' which formerly appeared in Article 5 is not considered to be of any significance. The provision applies unless the entry in the Appendix stating that 'clause 41B applies' is deleted, in which case disputes are to be determined by litigation under article 7B. Note that disputes relating to the enforcement of any decision of an adjudicator are excluded from arbitration and may therefore be referred to the court; however if either party, while complying with an adjudicator's decision, simply does not accept it as correct the original dispute is referable to arbitration. Disputes relating to the statutory tax deduction scheme and to VAT, which may only be determined by the bodies designated under the legislation relating to these matters, are excluded from arbitration.

All arbitrations under this Article are to be conducted under a version of the Construction Industry Model Arbitration Rules ('CIMAR' – see Appendix 2) adapted for use with JCT contracts and related sub-contracts (see later commentary).

Clause 41B

This clause now repeats elements of CIMAR relating to appointment of the arbitrator etc. It is therefore very much shorter than the equivalent clause 41 in previous editions of JCT 80 since much of the material now appears in CIMAR. Certain points should be noted.

- The previous prohibition, with certain exceptions, on reference of disputes to arbitration before actual or alleged practical completion, termination of the contractor's employment or abandonment of the works without written consent of both parties no longer appears. Any dispute arising under or in connection with the contract may therefore be referred to arbitration at any time.
- The previous clause 41.2 relating to joinder of related arbitration proceedings no longer appears. Joinder of arbitration proceedings is now covered by CIMAR Rule 3 and applies not only to disputes under main contracts and sub-contracts but also to disputes under consultants' agreements which also contain provision for CIMAR to apply, as is the case in the consultants'

agreements prepared by JCT for use where the main building contract is in a JCT Form. This should go a long way towards overcoming the problem caused by the mandatory enforcement of arbitration agreements under section 9 of the Arbitration Act 1996 and the fact that the court no longer has a discretion to refuse a stay of litigation proceedings in favour of arbitration where it used to be argued that the joinder of related disputes was only achievable by litigation.

- Clause 41B.2, which is in the same terms as clause 41.4 in previous editions, gives extensive powers to arbitrators to rectify the contract (a power now conferred in any case by section 48(5)(c) of the Arbitration Act 1996), to direct measurement and valuations and to open up, review and revise certificates, opinions, decisions, requirements and notices.
- As in the previous clause 41.6, clause 41B.4 contains the agreement of the parties to refer any question of law arising during the arbitration proceedings to the court and for either party to appeal to the court on any question of law arising out of an arbitral award. This obviates the need for a party to obtain the arbitrator's permission to refer a question of law under section 45(2)(b) or to obtain the leave of the court to refer a question of law to appeal under section 69(2)(b) of the Act.
- Footnote y.3 draws attention to the fact that special provisions apply where the site of the Works is in Scotland. Forms issued by the Scottish Building Contract Committee are available which incorporate the different law relating to arbitrations in Scotland.

Clause 41C

Where the reference to Clause 41B in the Appendix is deleted, and disputes are therefore to be referred to litigation, this clause seeks to give the courts similar power to open up, review and revise certificates, opinions, etc as an arbitrator would have had if Article 7A and clause 41B.2 had applied. The wording adopted is ingenious, but at the time of writing it still remains to be seen whether the courts will accept that a jurisdiction they do not inherently possess may be conferred on them by the provisions of a contract.

The JCT 1998 Edition of CIMAR

The Construction Industry Model Arbitration Rules are now incorporated into JCT contracts and related sub-contracts in their entirety together with the explanatory Notes issued by the Society of Construction Arbitrators. In addition the JCT have incorporated Supplemental and Advisory Procedures.

Mandatory procedures

There are only two additional mandatory procedures.

Rule 6.2

This Rule (laying down what the parties are to provide to the arbitrator after his appointment) is to be complied with by each party within 14 days after the arbitrator's acceptance of his appointment has been notified to the parties.

Rule 6.3

The procedural meeting under Rule 6.3 is to be convened by the arbitrator within 21 days of the date on which he has notified the parties of his acceptance of the appointment, unless within that time it has been decided in accordance with Rule 6.6 that a meeting is not required.

Advisory procedures

These procedures, which it must be noted are not mandatory and must be agreed between the parties after the arbitration has begun if they are to apply, basically lay down timescales for the procedures under Rules 7, 8 and 9 which, even then, will only apply unless other periods are ordered at the procedural meeting under Rule 6.3. They also lay down procedures to be followed in the event of non-compliance with the prescribed or ordered timescales.

The only other advisory procedure, subject to the same conditions, is an additional Rule 13.4.1 which provides that, where Rule 7 (Short Hearing) applies, notwithstanding Rules 7.5 and 7.7, each party shall bear its own costs and half the arbitrator's fees and expenses unless for special reasons the arbitrator at his discretion directs otherwise.

4.2 *The ICE Conditions of Contract*

The Institution of Civil Engineers issued amendments to the clauses dealing with the avoidance and settlement of disputes in the following forms of contract in March 1998 in order to take into account the Housing Grants, Construction and Regeneration Act 1996 (PART II)

> The ICE Conditions of Contract 5th Edition Clause 66
> The ICE Conditions of Contract 6th Edition Clause 66
> The ICE Design and Construct Conditions of Contract Clause 66
> The ICE Conditions of Contract Minor Works 2nd Edition. Clause 11 and Addendum A.

The wording of Clause 66 as revised is identical in the 5th and 6th Editions.

In the Design and Construct Conditions of Contract the word 'Engineer' which is used in the 5th and 6th Editions is replaced throughout by the words 'Employer's Representative'. In Clause 66(2)(a) the words 'Engineer's Representative' which are used in the 5th and 6th Editions are also replaced by the words 'Employer's Representative' and in the same Clause the word 'Engineer' is replaced by 'him'.

In the Minor Works Conditions of Contract the words 'Resident Engineer' appear in Clause A2(a) in lieu of 'Engineer's Representative'.

Otherwise the Clauses are identical.

We reproduce Clause 66 from the 5th and 6th Editions below.

Clause 66
AVOIDANCE AND SETTLEMENT OF DISPUTES

66 (1) In order to overcome where possible the causes of disputes and in those cases where disputes are likely still to arise to facilitate their clear definition and early resolution (whether by agreement or otherwise) the following procedure shall apply for the avoidance and settlement of disputes.

Avoidance of disputes

(2) If any any time

Matters of dissatisfaction

  (a) the Contractor is dissatisfied with any act or instruction of the Engineer's Representative or any other person responsible to the Engineer or

(b)   the Employer or the Contractor is dissatisfied with any decision opinion instruction direction certificate or valuation of the Engineer or with any other matter arising under or in connection with the Contract or the carrying out of the Works

the matter of dissatisfaction shall be referred to the Engineer who shall notify his written decision to the Employer and the Contractor within one month of the reference to him.

Disputes

(3)   The Employer and the Contractor agree that no matter shall constitute nor be said to give rise to a dispute unless and until in respect of that matter

(a)   the time for the giving of a decision by the Engineer on a matter of dissatisfaction under Clause 66(2) has expired or the decision given is unacceptable or has not been implemented and in consequence the Employer or the Contractor has served on the other and on the Engineer a notice in writing (hereinafter called the Notice of Dispute) or

(b)   an adjudicator has given a decision on a dispute under Clause 66(6) and the Employer or the Contractor is not giving effect to the decision, and in consequence the other has served on him and the Engineer a Notice of Dispute

and the dispute shall be that stated in the Notice of Dispute. For the purposes of all matters arising under or in connection with the Contract or the carrying out of the Works the word "dispute" shall be construed accordingly and shall include any difference.

(4)   (a)   Notwithstanding the existence of a dispute following the service of a Notice under Clause 66(3) and unless the Contract has already been determined or abandoned the Employer and the Contractor shall continue to perform their obligations.

(b)   The Employer and the Contractor shall give effect forthwith to every decision of

(i)   The Engineer on a matter of dissatisfaction give under Clause 66(2) and

(ii)   the adjudicator on a dispute given under Clause 66(6)

unless and until that decision is revised by agreement of the Employer and Contractor or pursuant to Clause 66.

Conciliation

(5)   (a)   The Employer or the Contractor may at any time before service of a Notice to Refer to arbitration under Clause 66(9) by notice in writing seek the agreement of the other for the dispute to be considered under the Institution of Civil Engineers' Conciliation Procedure (1994) or any amendment or modification thereof being in force at the date of such notice.

(b)   If the other party agrees to this procedure any recommendation of the conciliator shall be deemed to have been accepted as finally determining the dispute by agreement so that the matter is no longer in dispute unless a Notice of Adjudication under Clause 66(6) or a Notice to Refer to arbitration under Clause 66(9) has been served in respect of that dispute not later than 1 month after receipt of the recommendation by the dissenting party.

(6)   (a)   The Employer and the Contractor each has the right to refer any dispute as to a matter under the Contract for adjudication and either party may give notice in writing (hereinafter called the Notice of Adjudication) to the other at any time of his intention so to do., The adjudication shall be conducted under the Institution of Civil Engineers' Adjudication Procedure (1997) or any amendment or modification thereof being in force at the time of the said Notice. **Adjudication**

(b)   Unless the adjudicator has already been appointed he is to be appointed by a timetable with the object of securing his appointment and referral of the dispute to him within 7 days of such notice.

(c)   The adjudicator shall reach a decision within 28 days of referral or such longer period as is agreed by the parties after the dispute has been referred.

(d)   The adjudicator may extend the period of 28 days by up to 14 days with the consent of the party by whom the dispute was referred.

(e)   the adjudicator shall act impartially.

(f)   the adjudicator may take the initiative in ascertaining the facts and the law.

(7)   The decision of the adjudicator shall be binding until the dispute is finally determined by legal proceedings or by arbitration (if the contract provides for arbitration or the parties otherwise agree to arbitration) or by agreement.

(8)   The adjudicator is not liable for anything done or omitted in the discharge or purported discharge of his functions as adjudicator unless the act or omission is in bad faith and any employee or agent of the adjudicator is similarly not liable.

(9)   (a)   All disputes arising under or in connection with the Contract or the carrying out of the Works other than failure to give effect to a decision of an adjudicator shall be finally determined by reference to arbitration. The party seeking arbitration shall serve on the other party a notice in writing (called the Notice to Refer) to refer the dispute to arbitration. **Arbitration**

(b) where an adjudicator has given a decision under Clause 66(6) in respect of the particular dispute the Notice to Refer must be served within three months of the giving of the decision otherwise it shall be final as well as binding.

Appointment of arbitrator

(10) (a) The arbitrator shall be a person appointed by agreement of the parties.

President or Vice-President to act

(b) If the parties fail to appoint an arbitrator within one month of either party serving on the other party a notice in writing (hereinafter called the Notice to Concur) to concur in the appointment of an arbitrator the dispute shall be referred to a person to be appointed on the application of either party by the President for the time being of the Institution of Civil Engineers.

(c) If an arbitrator declines the appointment or after appointment is removed by order of a competent court or is incapable of acting or dies and the parties do not within one month of the vacancy arising fill the vacancy then either party may apply to the President for the time being of the Institution of Civil Engineers to appoint another arbitrator to fill the vacancy.

(d) In any case where the President for the time being of the Institution of Civil Engineers is not able to exercise the functions conferred on him by this clause the said functions shall be exercised on his behalf by a Vice-President for the time being of the said Institution.

Arbitration – procedure and powers

(11) (a) Any reference to arbitration under this Clause shall be deemed to be a submission to arbitration within the meaning of the Arbitration Act 1996 or any statutory re-enactment or amendment thereof for the time being in force. The reference shall be conducted in accordance with the procedure set out in the Appendix to the Form of Tender or any amendment or modification thereof being in force at the time of the appointment of the arbitrator. Such arbitrator shall have full power to open up review and revise any decision opinion instruction direction certificate or valuation of the Engineer or an adjudicator.

(b) Neither party shall be limited in the arbitration to the evidence or arguments put to the Engineer or to any adjudicator pursuant to Clause 66(2) or 66(6) respectively.

(c) The award of the arbitrator shall be binding on all parties.

(d) Unless the parties otherwise agree in writing any reference to arbitration may proceed notwithstanding that the Works are not then complete or alleged to be complete.

Witnesses

(12) (a) No decision opinion instruction direction certificate or valuation given by the Engineer shall disqualify him from being

called as a witness and giving evidence before a conciliator adjudicator or arbitrator on any matter whatsoever relevant to the dispute.

(b) All matters and information placed before a conciliator pursuant to a reference under sub-clause (5) of this clause shall be deemed to be submitted to him without prejudice and the conciliator shall not be called as witness by the parties or anyone claiming through them in connection with any adjudication arbitration or other legal proceedings arising out of or connected with any matter so referred to him.

Permission to reproduce material from the ICE Conditions of Contract has been granted by the copyright holder, the Institution of Civil Engineers. Copies of the Conditions can be obtained from Thomas Telford Publishing, 1 Heron Quay, London E14 4JD.

Two particular points are worthy of note in Clause 66. The first is that a failure to give effect to a decision of an adjudicator is specifically excluded from those matters that can be arbitrated, Clause 66(9)(a). The second is that the ICE have provided that unless a dispute that has been the subject of an adjudicator's decision is referred to arbitration within three months, that decision is final as well as binding. Should such a dispute be referred to arbitration after the three month period expires, any arbitrator appointed in respect of that dispute would not have any jurisdiction.

Table of Cases

Note:
The following abbreviations of law reports are used:

AC	Appeal Cases
All ER	All England Law Reports
BLR	Building Law Reports
Ch D	Law Reports (New Series) Chancery Division
CILL	Construction Industry Law Letter
Con LR	Construction Law Reports
Const LJ	Construction Law Journal
EG	Estates Gazette
EGLR	Estates Gazette Law Reports
KB	Law Reports, King's Bench
LJCh	Law journal Reports (New Series) Chancery
LJQB	Law Journal Reports (New Series) Queen's Bench
Lloyd's Rep	Lloyd's Law Reports
LT	Law Times Reports (New Series)
QB	Law Reports, Queen's Bench
TLR	Times Law Reports
WLR	Weekly Law Reports

Index

Index